SEA OF CORTEZ MAR
A Guide to the Common Fishe
Baja California to Panama

by

Daniel W. Gotshall

SEA CHALLENGERS
4 Sommerset Rise, Monterey, CA 93940

1998

A SEA CHALLENGERS PUBLICATION
AN IMPRINT OF SHORELINE PRESS

Distributed By
Pacific Books
P.O. Box 3562
Santa Barbara, CA 93130
(805) 687-8340
pacificbooks@gmail.com

Technical Editors: Fishers – Douglas Long
Invertebrates – Hans Bertsch

Copy Editor: Hans Bertsch

Forth Printing

Front Cover

Top photo: Clarion angclfish, *Holocanthus calarionensis*

Middle left photo: Gulf star, *Pentaceraster cumingi*

Middle right photo: Sedna, *Glossodoris sedna*

Bottom photo: Male Mexican hogfish, *Bodianus diplotaenia*

Library of Congress Cataloging-in-Publication Data

Gotshall, Daniel,
Sea of Cortez marine animals: a guide to common fishes and invertebrates – Baja California to Panama / by Daniel W. Gotshall.
p. cm.
Includes bibliographical references and index.
ISBN 10: 1-885375-17-4 and ISBN 13: 978-1-885375-17-9
1. Marine animals – Mexico-California, Gulf of – Identification.
2. Marine animals – Pacific Coast (Central America) – Identification
I. Title.
QL225.G676 1998
591.77'41-DC21

Shoreline Press
2573 Treasure Drive
Santa Barbara, CA 93105

Printed in China through Global Interprint, Santa Rosa, CA, USA
Typography and Prepress Production by Diana's Secretarial Service, Danville, CA, USA

DEDICATION

This book is dedicated to two men, both of whom were very influential in fostering my interest in the natural sciences and who served as the two key role models of my life.

Dedicated to Richard S. Croker, pioneering marine scientist and former diplomat to Mexico. 1908 - 1998

Dedicated to Donald Drake, teacher, mentor, inspiration and personal friend. 1915 - 1998

TABLE OF CONTENTS

ACKNOWLEDGEMENTS AND PHOTO CREDITS v
INTRODUCTION 1
PARTS OF A BONY FISH 4
PICTORIAL KEY TO FISH FAMILIES 5
SHARKS AND RAYS 13
MORAY, SNAKE, AND CONGER EELS 18
MACHETES, HERRINGS, LIZARDFISHES AND FROGFISHES 21
NEEDLEFISHES, AND SQUIRRELFISHES 22
CORNETFISH, TRUMPETFISH, AND SEAHORSES 24
SCORPIONFISH, SEA BASS AND SOAPFISH 25
BIGEYES, AND CARDINALFISHES 32
TILEFISHES, AND REMORAS 34
JACKS, AND ROOSTERFISH 35
SNAPPERS, MOJARRAS, GRUNTS, AND PORGYS 39
CROAKERS, GOATFISH, AND CHUBS 45
SPADEFISH 48
BUTTERFLYFISHES, ANGELFISHES, AND DAMSELFISHES 49
HAWKFISHES, MULLETS, AND BARRACUDAS 57
WRASSES, AND PARROTFISHES 58
JAWFISH, TRIPLEFIN, AND CLINID BLENNIES 66
TUBE, COMBTOOTH BLENNIES, AND GOBIES 68
SURGEONFISHES, MOORISH IDOLS, AND MACKERELS 71
FLOUNDERS, TRIGGERFISHES, AND FILEFISHES 74
TRUNKFISHES, PUFFERS AND PORCUPINEFISHES 77
PICTORIAL KEY TO INVERTEBRATES 81
SPONGES 83
HYDROIDS, ANEMONES, CORALS, AND GORGONIANS 84
POLYCHAETE WORMS 91
SCALLOPS, SNAILS, NUDIBRANCHS, AND OCTOPUS 92
LOBSTERS AND CRABS 95
SEA STARS, BRITTLE STARS, URCHINS, AND CUCUMBERS 97
BIBLIOGRAPHY 105
INDEX 106

ACKNOWLEDGEMENTS

Most field guides are the result of contributions by many scientists; this field guide is no exception. The following scientists contributed information on, or identification of the fishes: Jack Engle, William N. Eschmeyer, Ted Hobson, Robert N. Lea, Douglas Long, John McCosker, Theodore W. Pietsch, John Randall, Richard Rosenblat, and Robert Snodgrass. Information on and identification of some of the invertebrates was provided by Hans Bertsch, Jack Engle, Eric Hochberg Alex Kerstitch and Gary Williams.

Robert Schmieder allowed me access to the Rocas Alijos expedition photos. Ann Gotshall typed the original manuscript and helped edit the introduction.

The expeditions were conducted aboard the dive vessels Baja Explorador, Don Jose, Baja Treasure, Horizon and Megalodon; I owe a great debt of gratitude to the captains and crews of these vessels for their support and assistance.

Finally I wish to thank all of the divers who have participated in these expeditions in the Sea of Cortez to the Islas Revillegigado and outer Baja California. To all of the above and anyone I have failed to mention, my sincere thanks.

PHOTO AND ART CREDITS

Hans Bertsch	65 (invertebrates)
Mark Conlin	162
Steve Drogin	Backcover
D. Dvorak-Cordell Expeditions	18, 135
Art Haseltine	26
Ted Hobson	73
Alex Kerstitch	163
Kevin McDonnel	75
Diane R. Nelson	72
Doug Perrine	2
Marc Shargel	3, 4, 173, 174
Alan Studley	1, 13, 14
Tetsuo Uno	25, 29, 177
Marc Shargel	Computer enhanced photos for species 59, 78
Charles Turner (deceased)	95
All other photos by Daniel Gotshall	

Dawn Navarro produced the drawings for the following pictorial key to fish families: soapfish, seahorse, cornetfish, trumpetfish, squirrelfish, whale shark, snapper, grunt, angelfish, damselfish, hawkfish, parrotfish, jawfish, triplefin, clinid, and combtooth blennies, surgeonfish and trunkfish. Dawn Navarro also did the drawing of the Gulf grouper (#48).

Daniel J. Miller produced the rest of the drawings for the pictorial key to fish families.

Kim Odenweller was responsible for the drawings in the pictorial key to the invertebrates.

INTRODUCTION

In December 1977 my wife, Ann, and I travelled to La Paz, Baja California, Mexico, to explore the possibilities of doing a book on the fishes and invertebrates of the Sea of Cortez. As luck would have it we made contact with the new co-owners of the Baja Explorador.

This dive charter boat was scheduled to begin operations in the summer of 1978. Thus began a series of diving expeditions in the Sea of Cortez, as well as to the Islas Revillagigedo. The results of these initial expeditions were published in 1982 under the title "Marine Animals of Baja California." A second, revised and updated edition of this book was released in 1987.

Scope: This present field guide expands the number of species described in the previous book and increases the area of coverage from the waters off the outer coast of Baja California to Panama, including the offshore islands. In this field guide I describe 189 species of fish and 67 species of invertebrates. These are the most common fishes and invertebrates encountered by snorkelers and divers. I have not included most of the small cryptic fishes, rarely observed by divers, such as gobies, blennies, clingfish, sea catfish, brotulids and cuskeels, toadfish and most flatfishes. Only the most common invertebrates are described.

For more complete coverage of the fishes I recommend "Fishes of the Tropical Eastern Pacific" by Gerald R. Allen and D. Ross Robertson; for the invertebrates, "Sea of Cortez Marine Invertebrates" by Alex Kerstitch. I used both of these books extensively in researching this guide. For other books and field guides covering this region, please see the bibliography on page 105.

Conservation: When I first began diving in the Sea of Cortez, and around the Islas Revillagigedo in the late 1970s, large adults of most species of fishes were common. This was before the widespread use of monofilament gillnets, increased commercial fishing by foreign fleets and the increase in sport hook-and-line and spearfishing.

In the late 1970s and early 1980s it was unusual to share the night's anchorage with more than one or two boats. With the greater popularity of driving and flying to Baja California, and the corresponding increase in sportfishing activities in the 1980s, we began to witness the decrease in sizes, as well as the declines in numbers of many species of fish. These included hammerhead sharks, manta rays, mullet snapper, pargo, dog snappers, leopard groupers, in the Gulf of California, almaco jacks, leather bass, blue jacks and hammerhead sharks around Socorro and San Benedicto Islands, and in the Revillagigedos. All these fishes were either targeted or incidental catches of commercial and sport fishers. I am also sure that increased pollution from commercial and urban development had impacts on the Gulf's marine resources.

Fortunately, in the late 1980s and early 1990s, the Mexican government began setting aside marine areas and prohibiting all or most fishing activities. These protected marine areas include the Islas Revillagigedo, where all commercial fishing is prohibited, and Cabo Pulmo, the only coral reef system in the Gulf. More areas are under consideration. The Mexican government also curtailed many foreign commercial fishing activities. It is to be congratulated for these farsighted efforts to protect Mexico's marine resources.

Divers and snorkelers everywhere can participate in these efforts to protect the Sea of Cortez marine ecosystem, as well as all other marine ecosystems, by practicing non-intrusive, non-extractive activities.

These include the following:

1. The basic rule of "Take only pictures, leave only bubbles."
2. "Look, but do not touch" attitude toward all marine life.
3. By not treating marine animals as "pets," i.e., do not pet or feed! The same animal you train to take food from your hand today, may approach a spear fisherman tomorrow expecting to be fed.
4. Report all illegal activities.
5. When fishing, take only what you need and refrain from taking protected species such as cabrillo, lobsters and abalone; these latter invertebrates are reserved for the commercial cooperatives.

If we fail to protect this unique ecosystem now, the future of the Sea of Cortez, as well as the rest of coastal Mexico and Central America will not be able to provide stable fisheries in the future to feed the people of this region. This will undoubtedly result in a curtailment of sport diving and fishing, as well as a loss of opportunity for future generations of divers and snorkelers to enjoy the beauty and solitude of this underwater paradise.

Another way to help protect this area is by supporting organizations that are dedicated to protecting marine resources. One of the best for the Gulf area is Sea Watch, an organization founded by Mike McGettigan. This non-profit organization is dedicated to increasing the public's awareness of destructive activities in the Sea of Cortez and surrounding waters. They have been responsible for reporting illegal fishing activities to the Mexican Government, particularly activities around the Islas de Revillagigedo.

How to Use Field Guides: The most effective way to use this or any other field guide is to review, review, review. By constantly going over the photographs, you will find that when diving it is common to look at an animal and remember seeing it in the guide. You may not remember the name, but you should be able quickly to locate the animal in question in your field guide. I also believe that by constant review the shapes and colors of the animals become more familiar.

When using this guide, if an animal is unfamiliar, turn to the pictorial keys, page 5 for the fishes and page 83 for invertebrates. Look for the shape that most closely resembles the animal that you observed, then turn to that section and look at the photos. If your fish or invertebrate is not there, but you are convinced you have the right family of phylum, then you should consult other books or field guides (see Bibliography).

Common and Scientific Names: I have used, in most cases, the common and scientific names used in "The Reef Fishes of the Sea of Cortez." For fishes not included in that book, I have used the American Fisheries Society's "Common and Scientific Names of Fishes from the United States and Canada," and "Fishes of the Tropical Eastern Pacific." Spanish common names are from "The Reef Fishes of the Sea of Cortez." Douglas Long and Hans Bertsch also provided spanish names for fishes that were not included in that book. Few of the invertebrates have common names; thus, for descriptive purposes, I have either coined names or used the names in "Sea of Cortez Marine Invertebrates."

New Information: Our knowledge of the fishes and invertebrates of this region continues to grow as more researchers and naturalists continue to describe new species and make new observations on known species. You can participate in this effort by reporting, preferably with a photograph, any species that you cannot identify, to scientists working in this region. Such scientists are also interested in obtaining new depth and geographic range data for the animals they are studing. If you are unable to locate someone to help you, please send all such relevant information to Sea Challengers, 4 Sommerset Rise, Monterey, CA 93940, or email SEACHALL@AOL.COM, and we will pass it on to the scientist best qualified to help you identify the animal in question, or to use any new information you may send. If such information is used in future editions of this book, you will be acknowledged for your contribution.

PARTS OF A BONY FISH

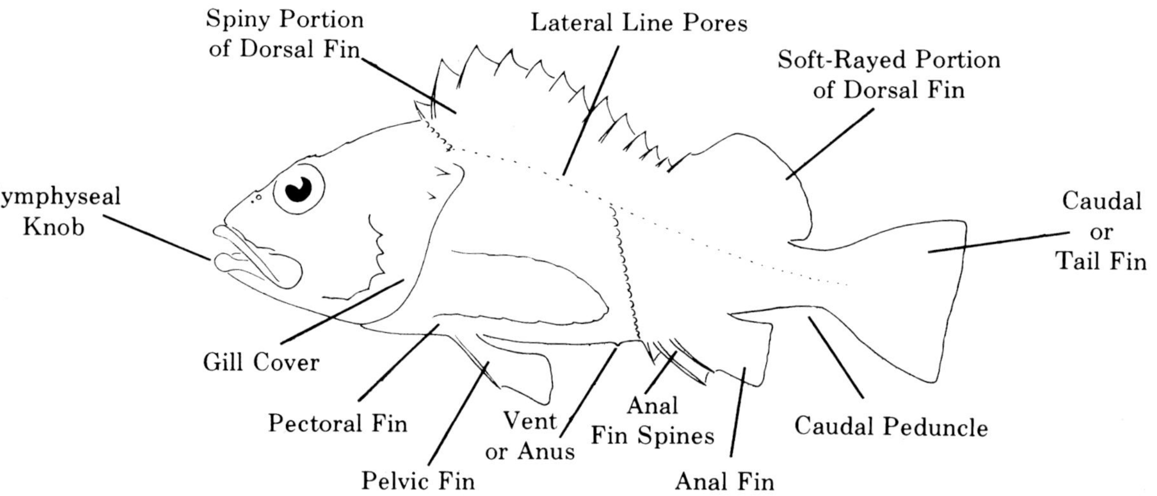

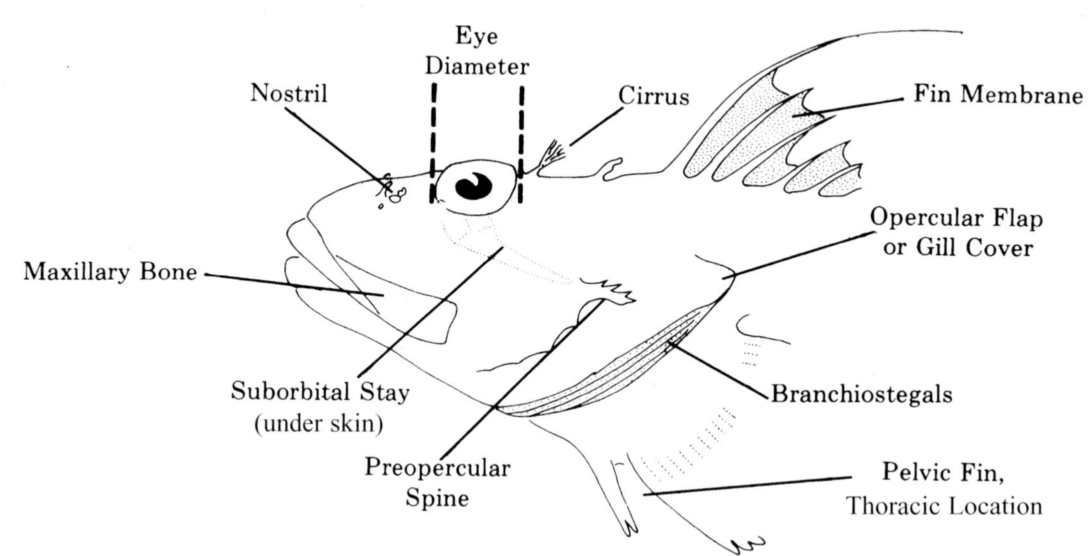

Parts of a Bony Fish

(Drawings by Daniel J. Miller)

PICTORIAL KEY TO FISH FAMILIES

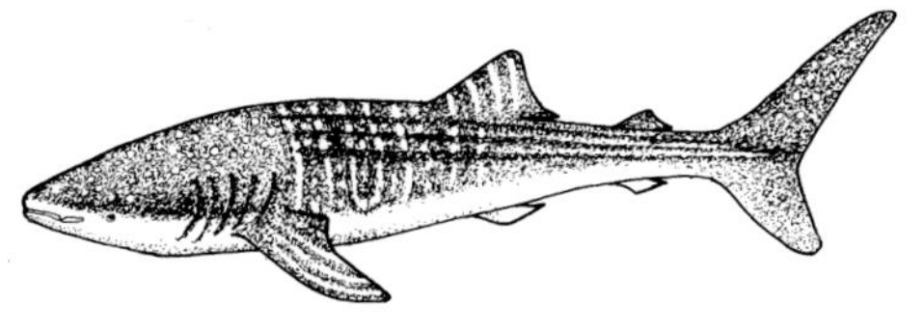

WHALE SHARK P. 13

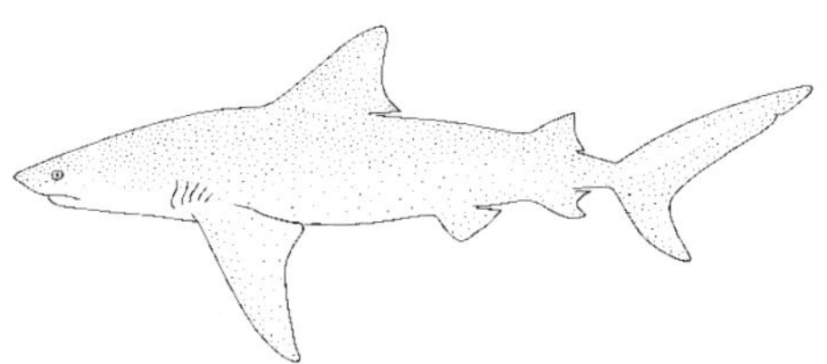

REQUIEM SHARKS P. 13

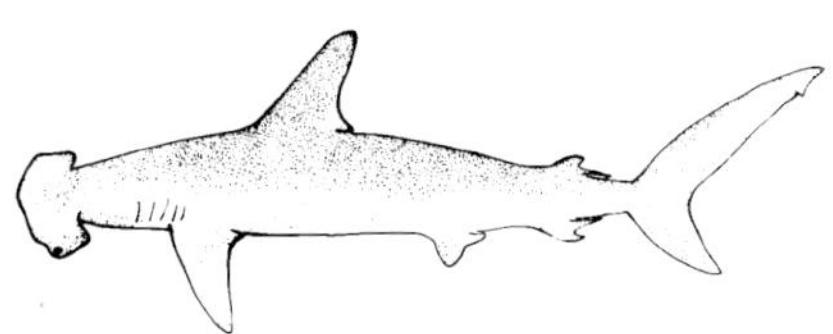

HAMMERHEAD SHARKS P. 14

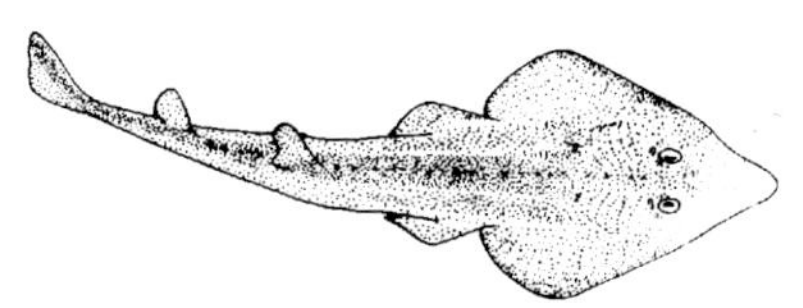

GUITARFISHES P. 14

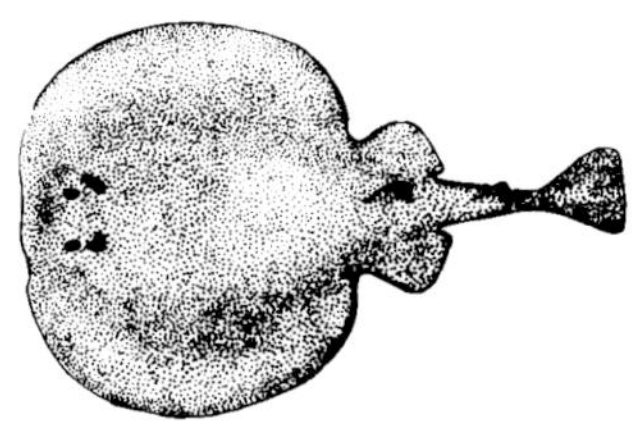

ELECTRIC RAYS P. 15

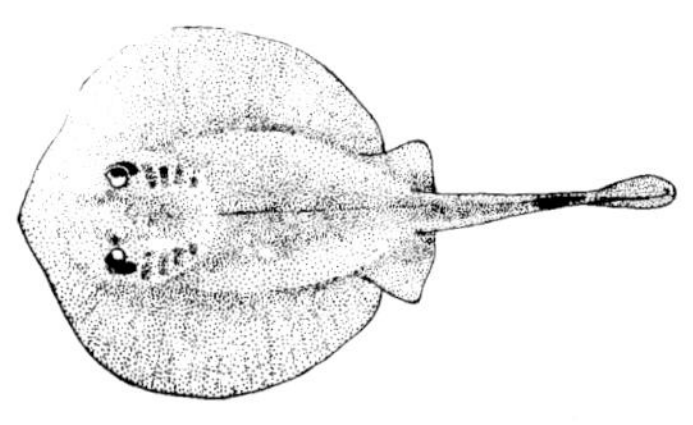

STING RAYS P. 16

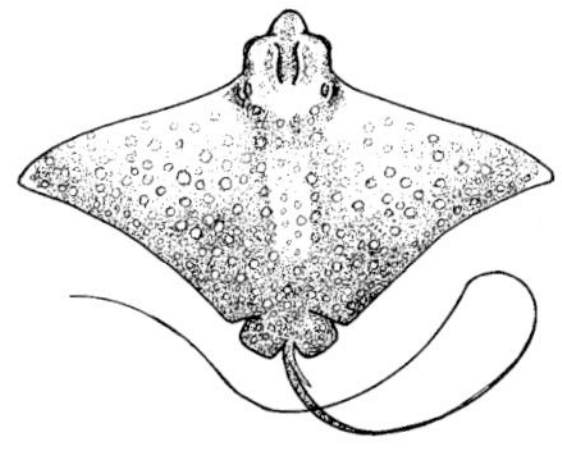

EAGLE RAYS P. 17

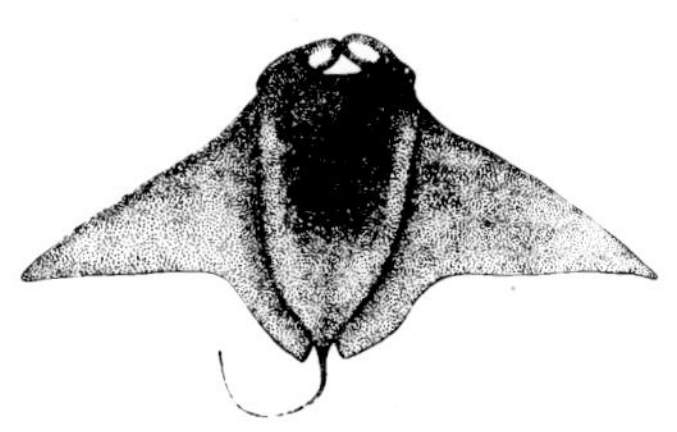

MANTA RAYS P. 17

PICTORIAL KEY TO FISH FAMILIES

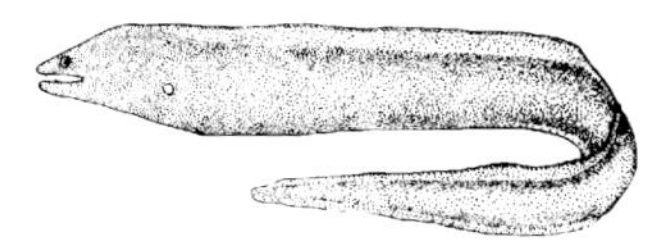

MORAYS P. 18

SNAKE EELS P. 20

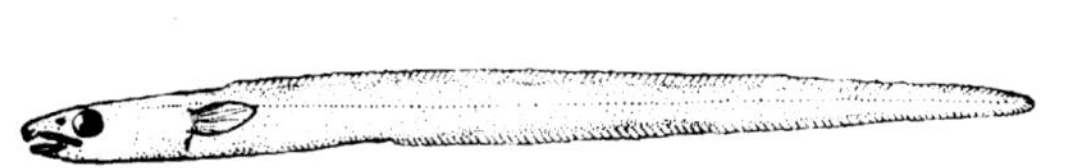

CONGER EELS P. 21

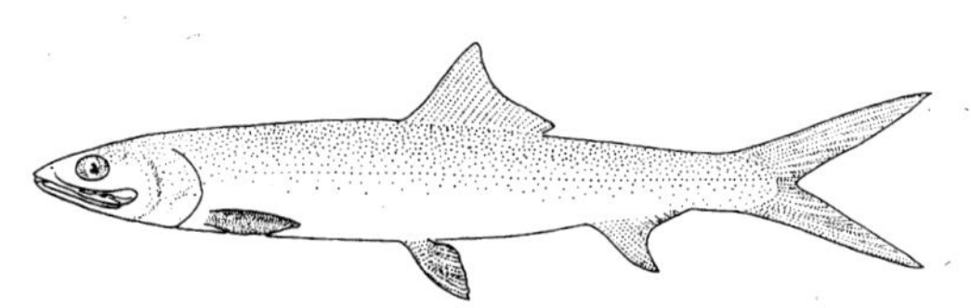

MACHETES P. 21

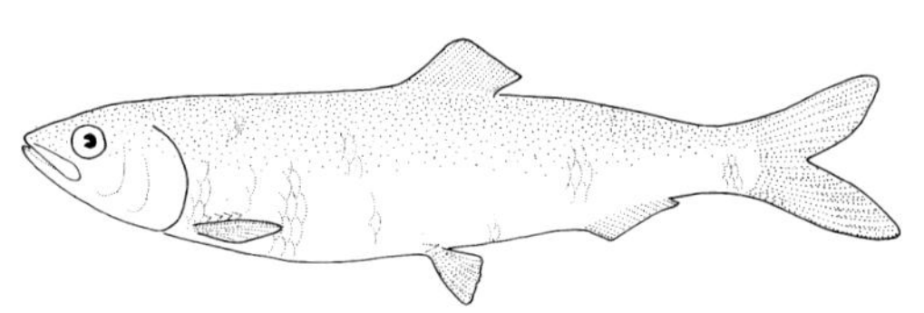

HERRINGS P. 21

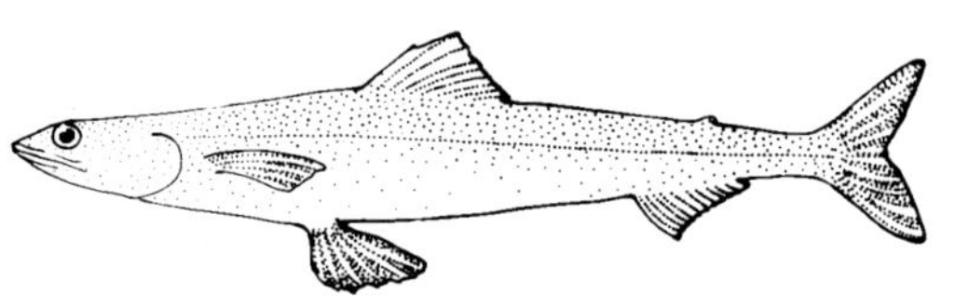

LIZARDFISHES P. 22

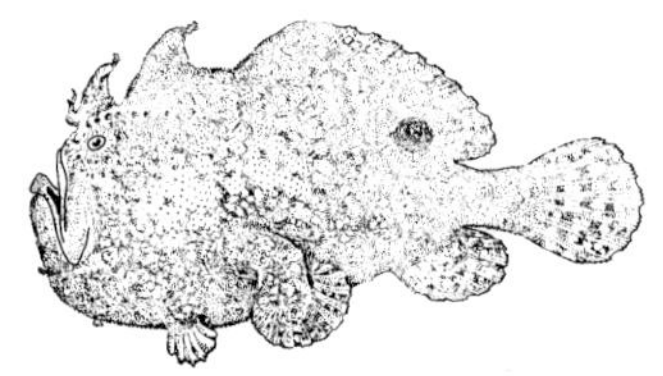

FROGFISHES P. 22

NEEDLEFISHES P. 22

PICTORIAL KEY TO FISH FAMILIES

SQUIRRELFISHES P. 23

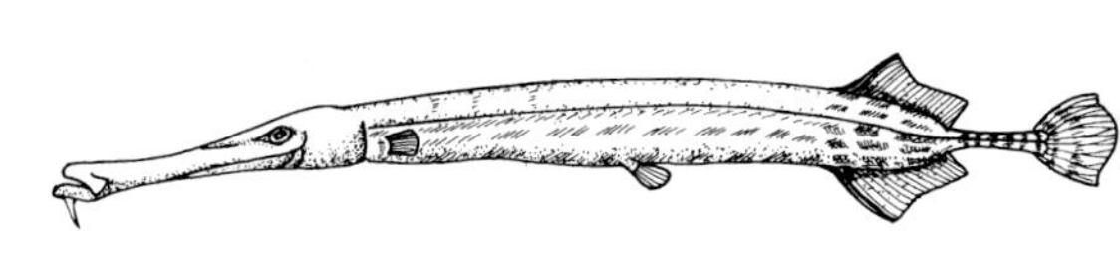

TRUMPETFISHES P. 24

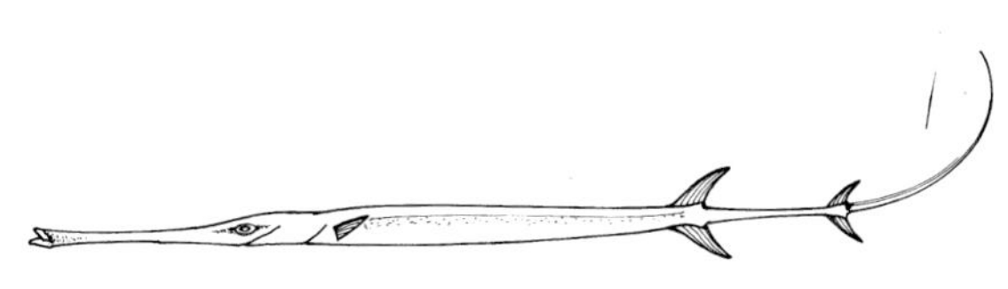

CORNETFISHES P. 24

SEAHORSES P. 25

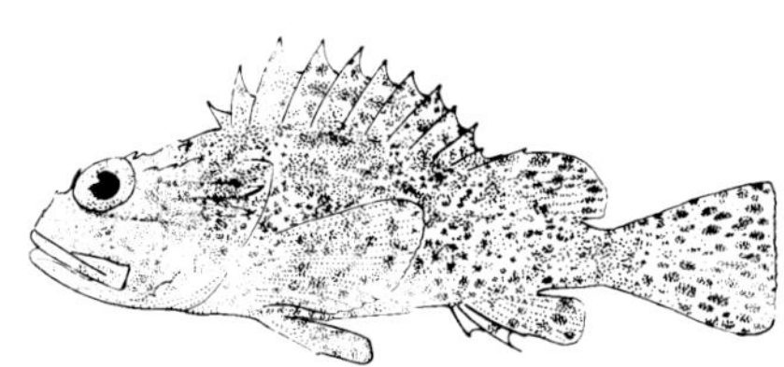

SCORPIONFISHES P. 25

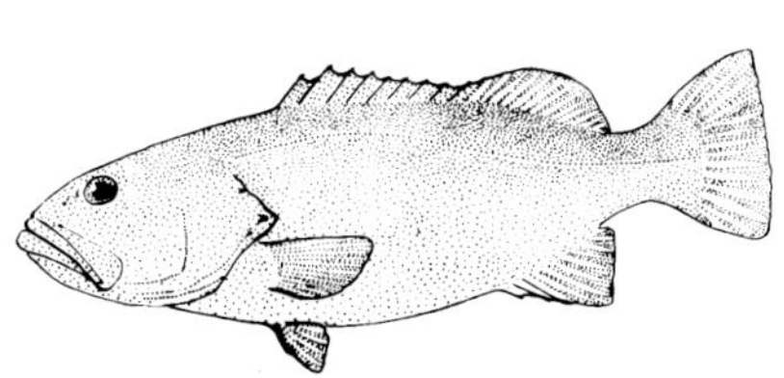

SEA BASSES P. 26

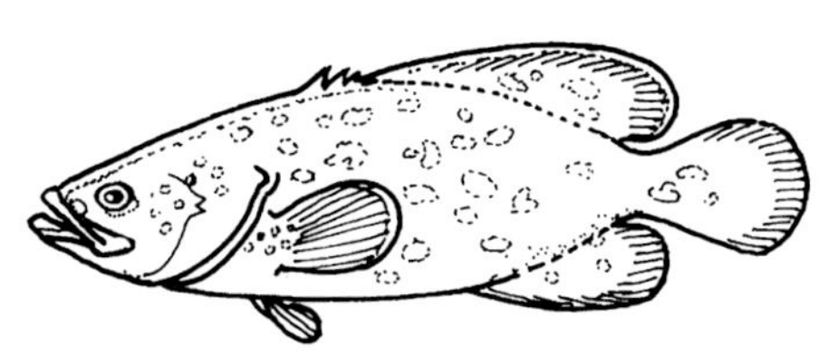

SOAPFISHES P. 32

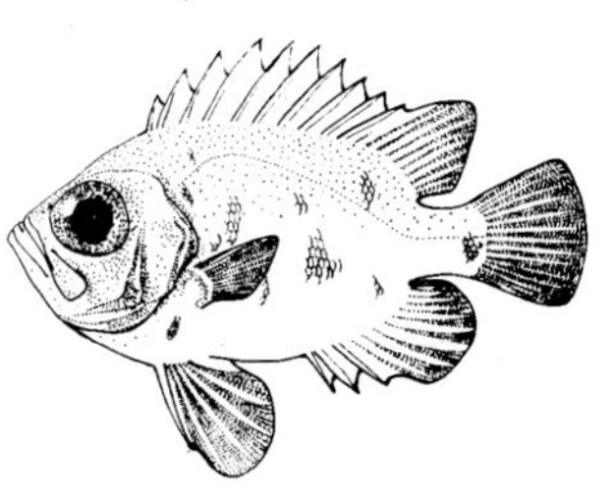

BIGEYES P. 32

PICTORIAL KEY TO FISH FAMILIES

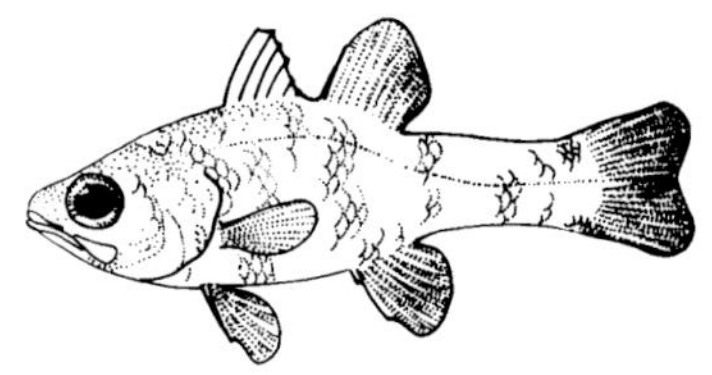

CARDINALFISHES P. 33

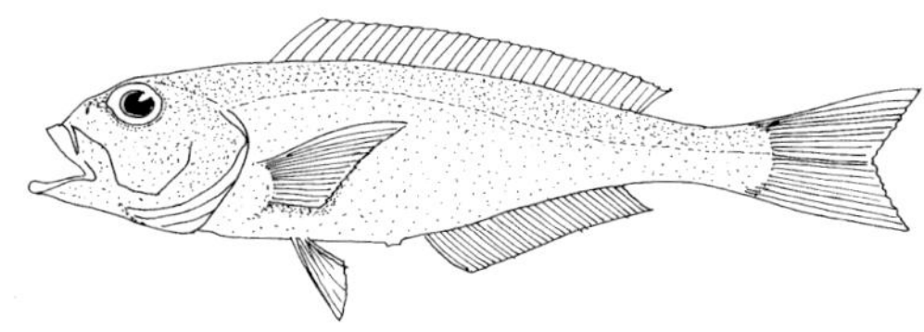

TILEFISHES P. 34

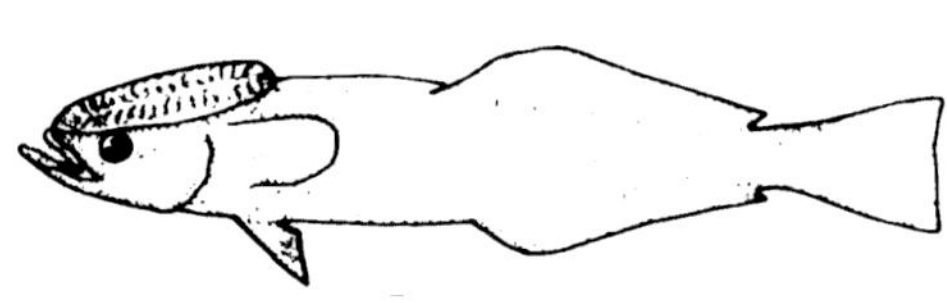

REMORRAS P. 35

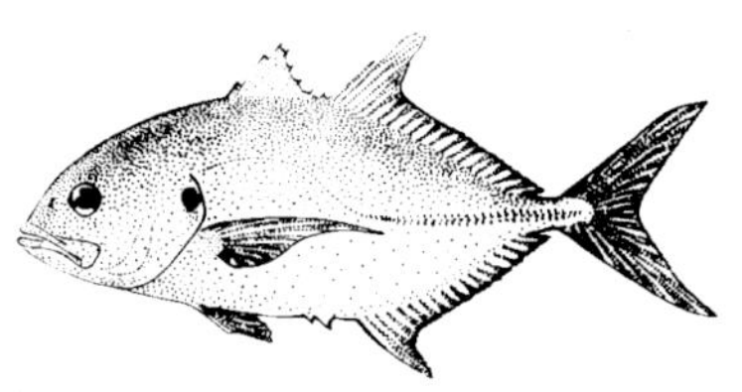

JACKS P. 35

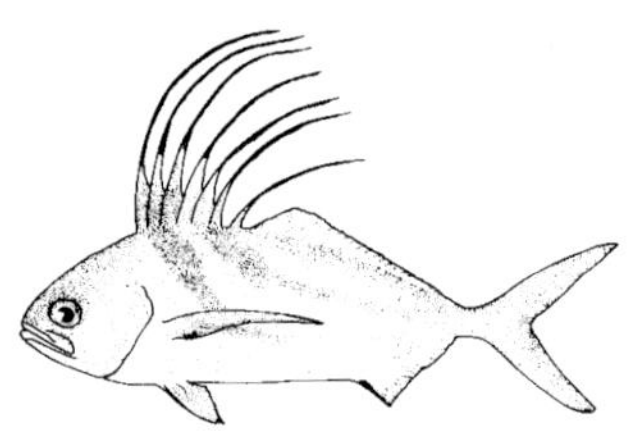

ROOSTERFISH P. 39

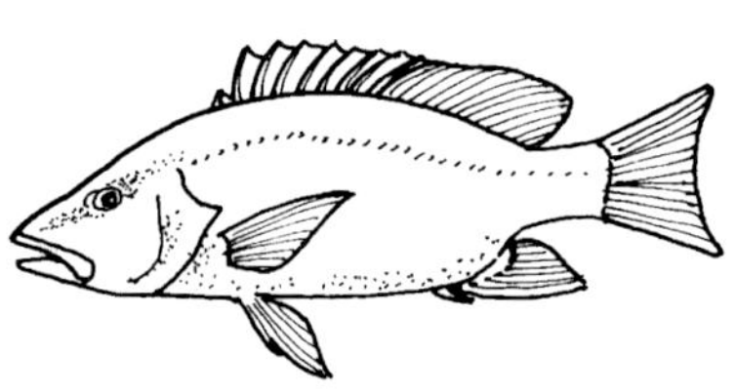

SNAPPERS P. 39

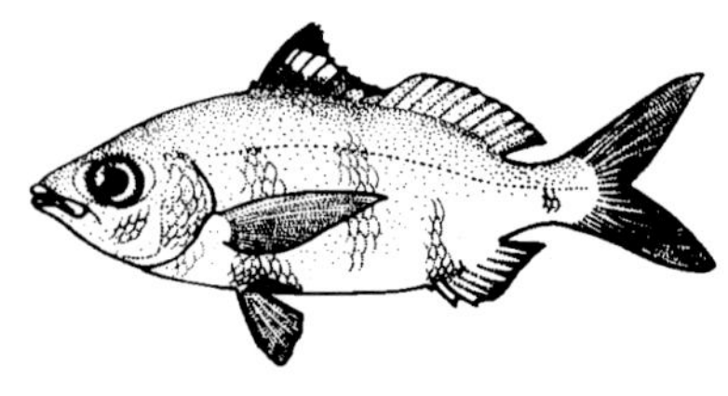

MOJARRAS P. 42

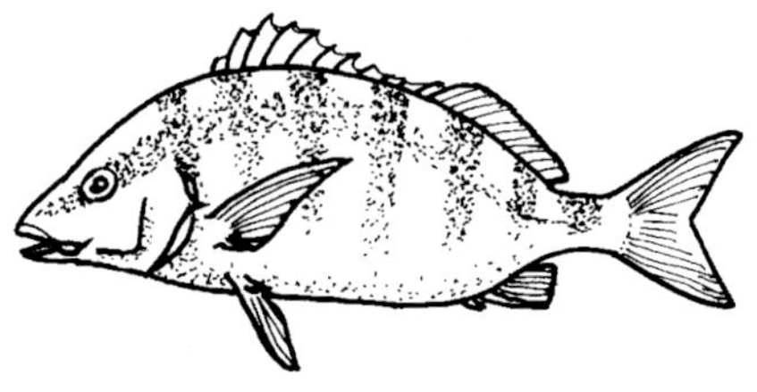

GRUNTS P. 42

PICTORIAL KEY TO FISH FAMILIES

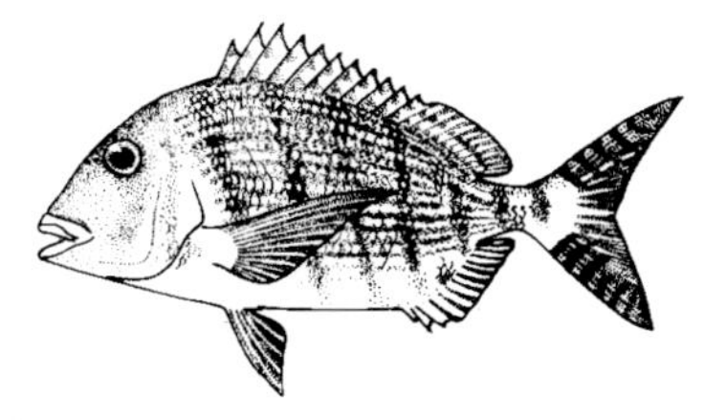

PORGIES P. 45

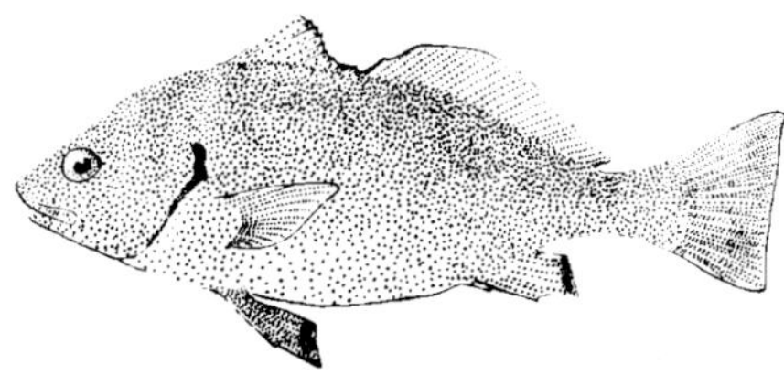

DRUMS AND CROAKERS P. 45

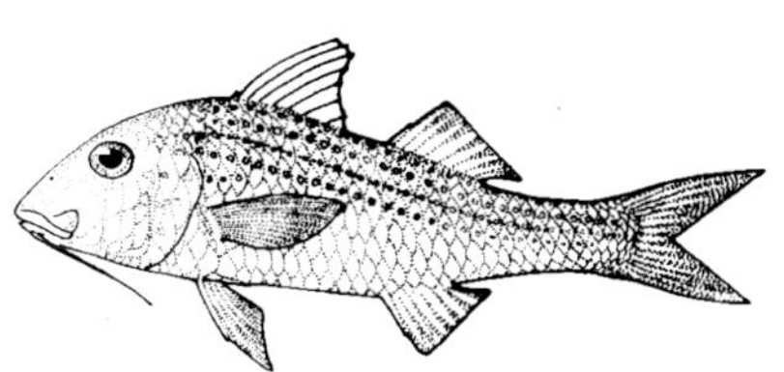

GOATFISHES P. 46

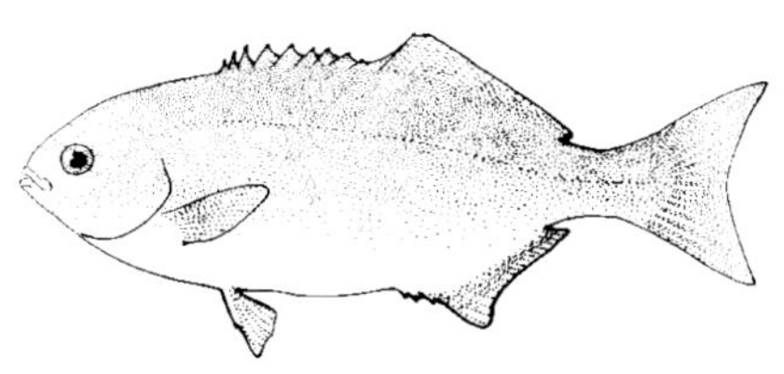

SEA CHUBS P. 46

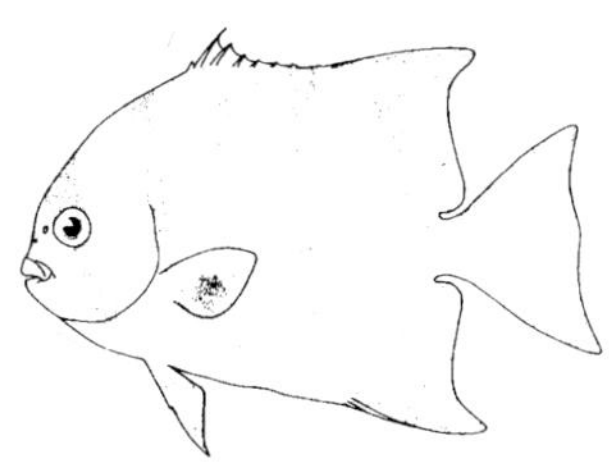

SPADEFISHES P. 48

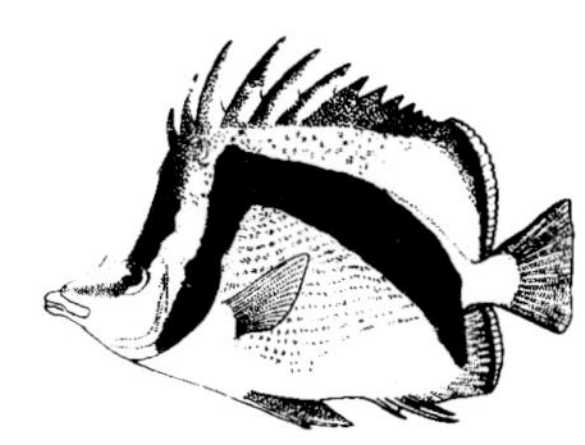

BUTTERFLYFISHES P. 449

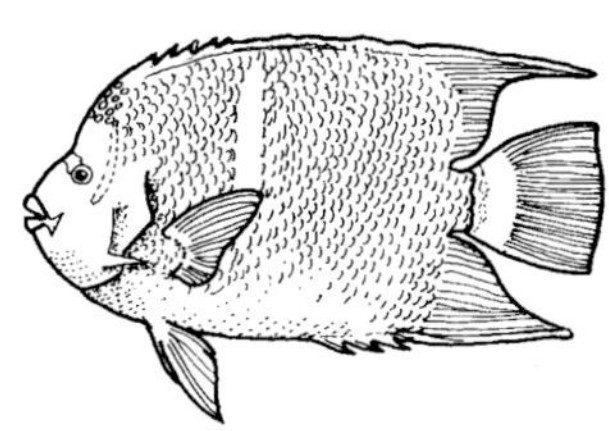

ANGELFISHES P. 50

DAMSELFISHES P. 52

PICTORIAL KEY TO FISH FAMILIES

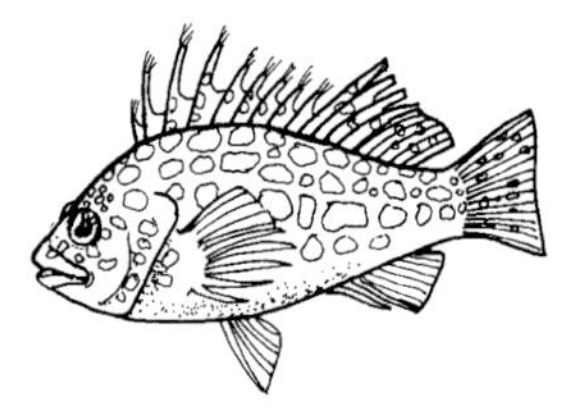

HAWKFISHES P. 57

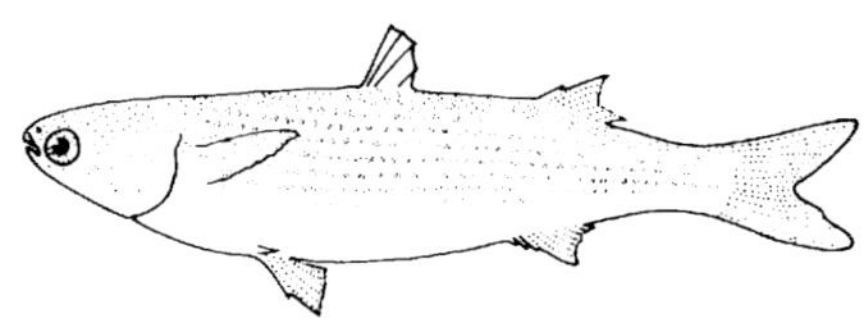

MULLETS P. 58

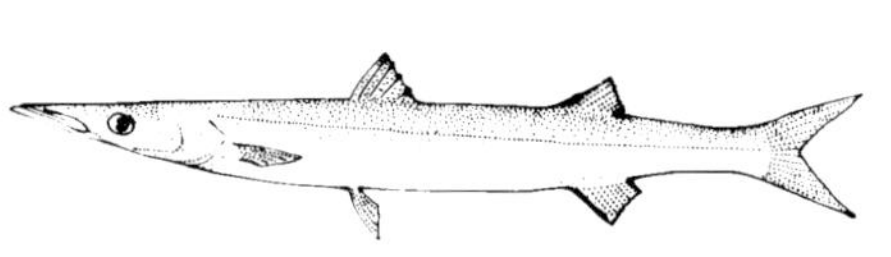

BARRACUDAS P. 58

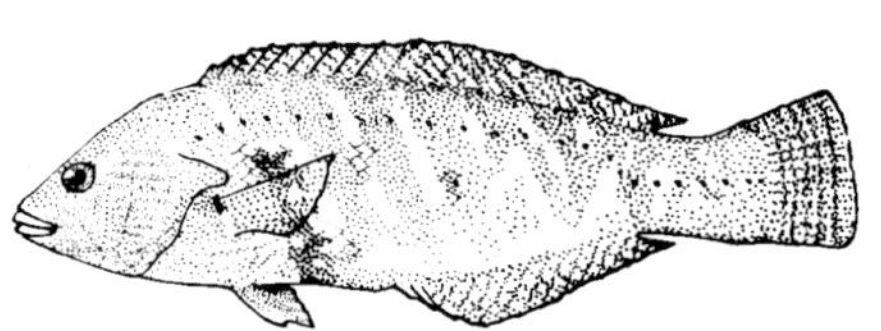

WRASSES P. 58

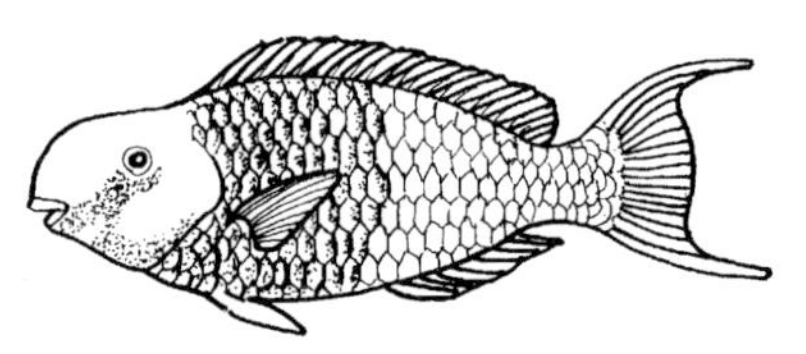

PARROTFISHES P. 64

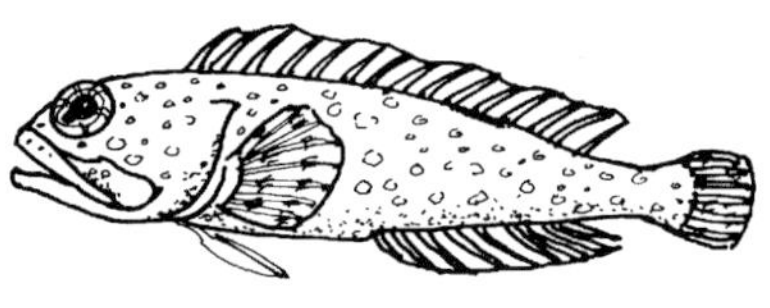

JAWFISHES P. 66

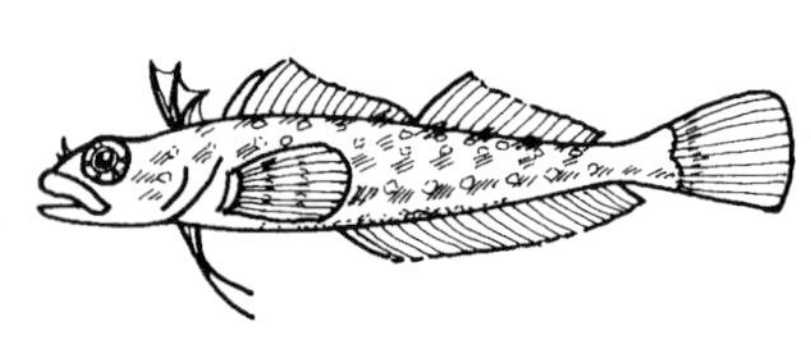

TRIPLEFIN BLENNIES P. 67

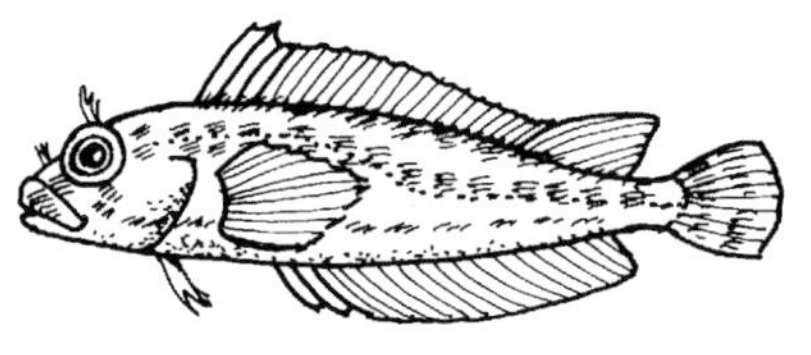

CLINID AND TUBE BLENNIES P. 67

PICTORIAL KEY TO FISH FAMILIES

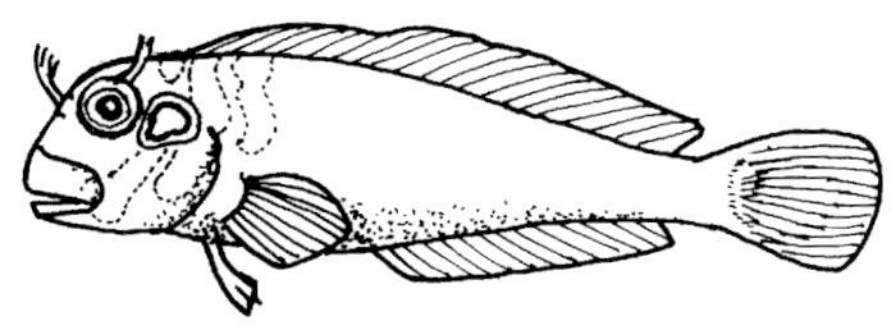

COMBTOOTH BLENNIES P. 69

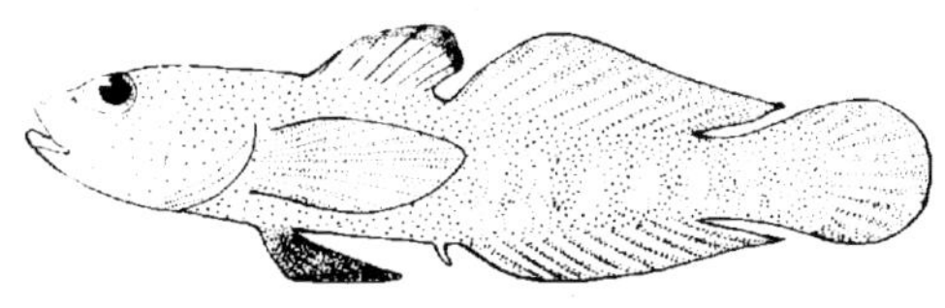

GOBIES P. 69

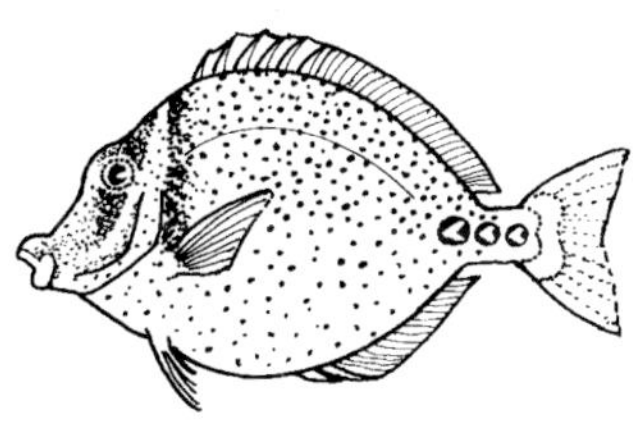

SURGEONFISHES & MOORISH IDOLS P. 71

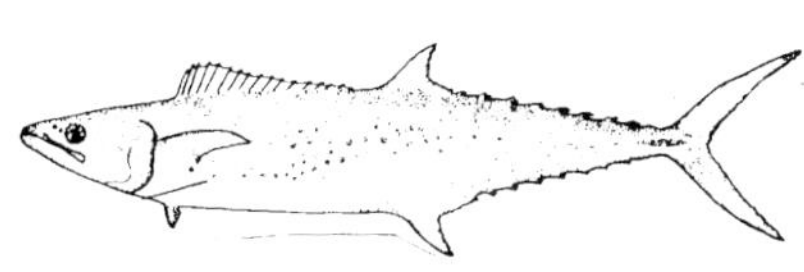

MACKERELS P. 74

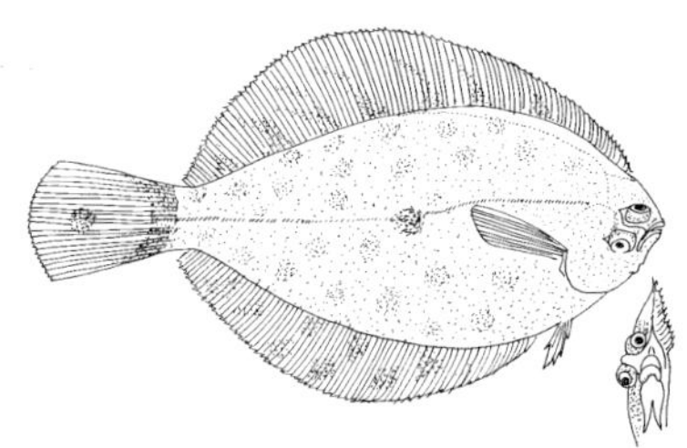

FLOUNDERS P. 74

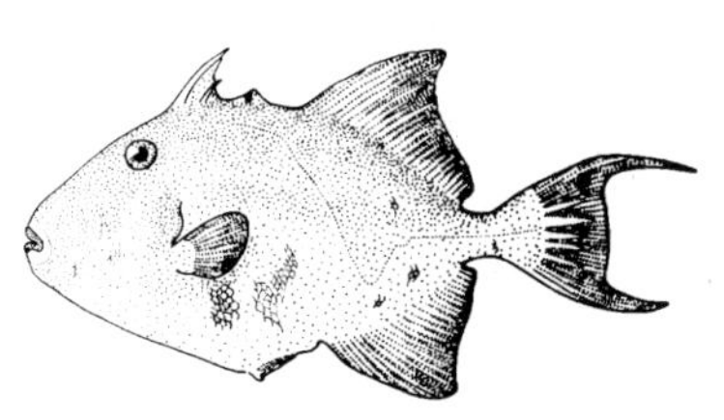

TRIGGERFISHES & FILEFISHES P. 75

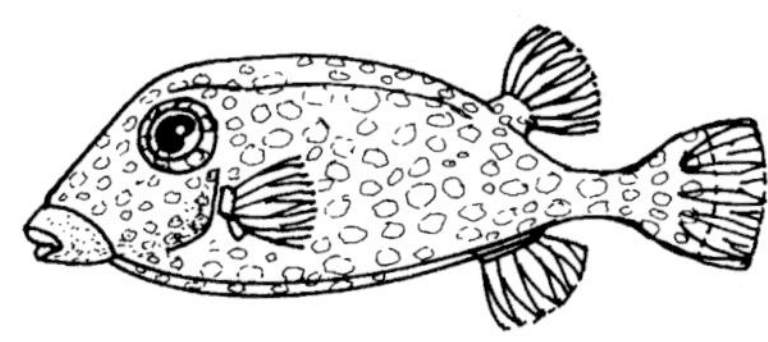

TRUNKFISHES P. 77

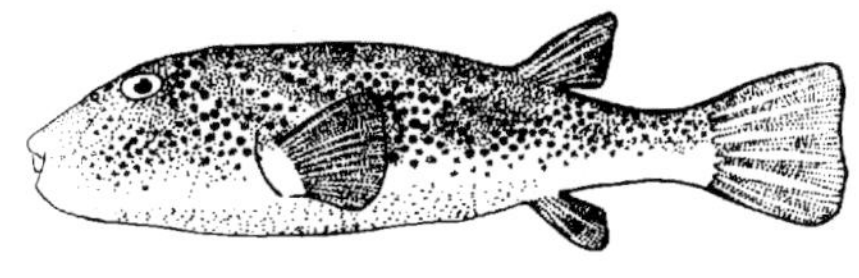

PUFFERS P. 77

PICTORIAL KEY TO FISH FAMILIES

PORCUPINEFISHES P. 79

FAMILY RHINCODONTIDAE

Whale Shark

1. **WHALE SHARK (Tiburon ballena)** ***Rhincodon typus***

Identification: The prominent ridges on the sides of the blackish body and the white spots, stripes and bars on the upper body are very distinctive. *Size*: Length to about 60 feet (18 m). *Habitat*: Surface waters where concentrations of plankton occur. *Range*: Worldwide in warm tropical seas, on this coast from San Diego, California, to Chile. *Natural History*: Little is known of their reproductive habits, but scientists believe that the eggs develop in cases within the female. Their food consists of planktonic fishes, crustaceans and squids.

A. Studley

FAMILY CARCHARHINIDAE

Requiem Sharks

2a. **SILKY SHARK (Tiburon jaqueton)** ***Carcharhinus falciformis***

2b. **GALAPAGOS SHARK (Tiburon de Galapagos)** ***Carcharhinus galapagensis***

Identification: The silky shark has a small first dorsal fin with a rounded tip. The second dorsal fin has a very long free tip. Color ranges from brown to blue-grey. The look-alike Galapagos shark, *C. galapagensis,* has a large dorsal fin that is pointed and the tips of most fins are dusky. These are the two most common sharks encountered by divers that lack other distinctive features. *Size*: Silky shark: Length to about 7.5 ft. (230 cm). Galapagos shark: Length to about 12 ft. (350 cm). *Habitat*: Silky sharks: Oceanic waters as well as nearshore. In depths of 60 to 1600 ft. (18 to 500 m). Galapagos shark: Usually around oceanic islands. *Range*: Silky shark: All of the tropical oceans; on this coast from Gulf of California to Chile. Galapagos shark: Tropical. Occurs worldwide; on this coast from Baja California to Peru. *Natural History*: Both sharks feed on fishes. Silky sharks produce 2 to 14 pups, Galapagos sharks 6 to 14 pups per litter. The Galapagos shark has attacked humans.

Silky Shark D. Perrine Galapagos Shark M. Shargel

3. WHITETIP REEF SHARK (Tiburon coraleno nato)

Triaenodon obesus

Identification: This slender shark has white tips on the first dorsal fin and upper lobe of the tail; the second dorsal fin is about two thirds the height of the first dorsal fin. *Size*: Length to 6.9 ft. (2.1 m). *Habitat*: Nocturnal; during daylight these sharks are usually found in caves or under ledges. *Range*: Indo-Pacific; in the eastern Pacific as far north as Islas Revillagigedo to Panama and Islas Galapagos. *Natural History*: A litter can consist of as many as 5 sharks. A nocturnal predator on fishes, crabs, lobsters and octopuses.

FAMILY SPHYRNIDAE
Hammerhead Sharks

4. SCALLOPED HAMMERHEAD

(pez martillo) ***Sphyrna lewini***

Identification: The four lobes on the front margin of the head are the best field characters. *Size*: Length to 13.9 ft. (4.2 m). *Habitat*: In surface waters around offshore reefs, sea mounts, and islands. *Range*: Worldwide in tropical oceans. Southern California to Ecuador on this coast; Southern California to Ecuador. *Natural History:* The most abundant of the hammerhead sharks. They feed on fishes, crustaceans, octopuses and even sea snakes on occasion. A litter may consist of 15 to 31 young sharks.

M. Shargel

FAMILY RHINOBATIDAE
Guitarfishes

5. SHOVELNOSE GUITARFISH

(pez guitarra) ***Rhinobatus productus***

Identification: The triangular-shaped head (disc), long-pointed snout and lack of spots or other markings will distinguish this shark from the banded guitarfish (# 6). *Size*: Length to 67 inches (170 cm) and 40 lbs (18.4 kg). *Habitat*: Usually on sandy bottoms, near reefs and in bays and estuaries, to depths of about 60 ft. 920 m). *Range*: San Francisco, California, to the Gulf of California. *Natural History*: These sharks bear live young, as many as 28 per litter. Shovelnose guitarfish feed on crabs, clams, worms and even small fishes.

6. BANDED GUITARFISH
(guitarra rayada) ***Zapteryx exasperata***

Identification: The disc, which is about as wide as long in adults, and many dark bands across the back separate this guitarfish from the shovelnose guitarfish, *Rhinobatus productus*. *Size*: Length to 5.1 ft. (1.6 m). *Habitat*: Soft bottoms around reefs; in caves in shallow water. *Range*: Newport Beach, California, to Ecuador.

FAMILY TORPEDINIDAE
Electric Rays

7. GIANT ELECTRIC RAY
(raya electrica) ***Narcine entemedor***

Identification: Differs from bullseye electric ray by possessing several large spots on back. However these spots may be absent on some individuals. *Size*: Length to 3 ft. (0.9 m). *Habitat*: Shallow, soft bottoms around reefs and in bays. *Range*: Gulf of California and Bahia Sebastian Vizcaino, Baja California, to Panama. *Natural History*: This nocturnal predator feeds on various types of worms.

8. BULLSEYE ELECTRIC RAY
(ray electrica ocelada)

Diplobatis ommata

Identification: Markings on the bullseye's back are very distinctive. *Size*: Length to about 1.5 ft. (0.5 cm). *Habitat*: Shallow waters to depths of 211 ft. (64 m). *Range*: Gulf of California to Ecuador. *Natural History*: The prey of this night feeder consists of small crustaceans and worms.

FAMILY DASYATIDAE
Stingrays

9. DIAMOND STINGRAY (raya de espina) ***Dasyatis brevis***

Identification: The diamond- or rhomboid-shaped body and short tail, less than 1.5 times the length of the body, and the sting located closer to the base of the tail than the tip, are good field characters. *Size*: Width to 5.8 ft. (1.8 m). *Habitat*: Soft bottoms around reefs and in mangroves to depths of about 60 ft. (18 m). *Range*: Central California to Ecuador and offshore islands. *Natural History*: These rays feed primarily on crabs.

FAMILY UROLOPHIDAE
Round Stingrays

10. BULLSEYE STINGRAY (raya espina) ***Urolophus concentricus***

Identification: Differs from Cortez round stingray in possessing dark markings that form concentric rings on outer edge of the disc; lacks black spots. *Size*: Width to about 12 inches (0.3 m) length to about 20 inches (0.5 m). *Habitat*: Found on shallow, sandy areas around reefs and in bays. Range: Gulf of California. *Natural History*: These rays feed on crabs and worms.

11. ROUND STINGRAY (raya redonda) ***Urolophus halleri***

Identification: The pointed snout of this round stingray protrudes slightly from the disc. The short, stout tail has one dorsal fin. Usually tan or light brown in color, with the central portion of the disc having large brown circular markings superimposed on dense brown spots and vermiculations. *Size*: Length to about 22 inches (55 cm). Width of disc to about 12 inches (31 cm). *Habitat*: Usually on soft bottoms, sand or mud in bays, estuaries and other nearshore waters. Shallow waters to about 70 ft. (21 m). *Range*: Eureka, California, to Gulf of California and Panama. *Natural History*: Round stingrays bear up to 6 young. Mating occurs in June, and they give birth in September. These very common stingrays feed on small invertebrates and fishes. They are responsible for most of the stingray wounds suffered by swimmers and waders.

12. **CORTEZ ROUND STINGRAY**
(raya manchada) ***Urolophus maculatus***

Identification: Distinguished by a brownish disc covered with large black spots and light buff circular blotches. *Size*: Length to 16.5 inches (42 cm). *Habitat*: Shallow, sandy bottoms around islands and in bays. *Range*: Gulf of California, and outer coast of Baja California from Bahia Magdalena south. Depth range intertidal to about 6.6 ft. (20 m). *Natural History*: These rays feed on worms and crustaceans during the daylight hours.

A. Studley

FAMILY MYLIOBATIDAE
Eagle Rays

13. **SPOTTED EAGLE RAY**
(chucho pintado) ***Aetobatus narinari***

Identification: The rounded snout, triangular wings and upper body, covered with large white spots readily identify this ray. The under body is white. *Size*: Width of disc to about 8 ft. (2.5 m). *Habitat*: Around reefs and in estuaries. *Range*: Worldwide in tropical and semitropical seas. On this coast, they may be found from southern California and the Gulf of California south to Chile. *Natural History*: These distinctive rays feed on clams, oysters and other bivalves. A female will give birth to one to four young.

FAMILY MOBULIDAE
Manta Rays

14. **MANTA RAYS (mantaraya)**
Manta birostris

Identification: The long lobes on either side of head, short tail (about equal to or shorter than body length), and mouth located in front of head, not underside, separate this pelagic plankton feeder from the smaller, long tailed (about twice body length) mobulas, not illustrated (NI). *Size*: Width to 25 ft. (7.6 m). *Habitat*: Pelagic, often feeds around offshore reefs and rocks. *Range*: Worldwide in tropical waters; on this coast from Santa Barbara Island, California, to Peru. *Natural History*: Manta rays feed on planktonic crustaceans and even small fishes. They are often hosts to remoras. These large docile fish make use of cleaners, such as Clarion angelfish.

FAMILY MURAENIDAE
Morays

15. ZEBRA MORAY (morena cebra)
Gymnomuraena zebra

Identification: The members of this genus have blunt, molarlike teeth instead of sharp, canine teeth. The only moray in the Gulf with white bands on a reddish brown to black body. *Size*: Length to about 4 ft. (150 cm). *Habitat*: Crevices and holes, rocky reefs; occasionally found on sand bottoms. *Range*: Indo-Pacific and from mid-Gulf of California to Peru and some of the offshore islands.

16. PANAMIC GREEN MORAY (morena verde)
Gymnothorax castaneus

Identification: Posterior nostrils lack tubes; dorsal fin well developed; greenish brown, occasionally with small, white spots. *Size*: Length to over 4 ft. (1.2 m). *Habitat*: Crevices and holes during day; at night may be seen any place on reef or over sand or mud bottoms when foraging. *Range*: Rocas Alijos, Baja California, and Gulf of California to Panama and Islas Galapagos. *Natural History*: These very common morays feed on a variety of fishes and crustaceans at night as well as during the day.

17. FINE-SPOTTED MORAY (morena pintita) ***Gymnothorax dovi***

Identification: This brown or black moray is easily identified by the numerous small white dots on the head and upper body. *Size*: Length to about 4.5 ft. (140 cm). *Habitat*: In reef crevices. *Range*: Central Gulf of California to Ecuador and offshore islands. *Natural History*: Like other large morays, fine-spotted morays feed on small fishes and crustaceans. Foraging takes place at night as well as during daylight.

18. ARGUS MORAY (morena pecas blancas)

Muraena argus

Identification: The brown body has three rows of yellow blotches and is covered with white spots which are more numerous on the head. There is a black spot around the gill opening. *Size*: Length to about 47 inches (120 cm). *Habitat*: Rocky crevices and caves. *Range*: Rocas Alijos, Baja California, to Peru and the Islas Galapagos.

D. Dvorak - Cordell Expeditions

19. HOURGLASS MORAY (morena de piedra)

Muraena clepsydra

Identification: The oscillated black spot that covers the gill opening and the large white spot on the rear of the lower jaw, as well as the cream colored spots on the brown body are distinctive. *Size*: Length to about 3 ft. (100 cm). *Habitat*: Crevices in rock and coral reefs. *Range*: Central Gulf of California to Ecuador and the Islas Galapagos.

20. JEWEL MORAY (morena pinta)

Muraena lentiginosa

Identification: Both the anterior and posterior nostrils are tubed in this genus; the chainlike rows of light spots ringed by dark brown halos are very distinctive on this nocturnal forager. *Size:* Length to 2 ft. (0.6 m). *Habitat*: Crevices in reef during the day; anywhere on reef at night when foraging. *Range*: Central Gulf of California to Peru and most offshore islands.

21. **TIGER REEF EEL**
(morena atigrada) ***Scuticaria tigrina***
Identification: Like all morays, this eel lacks pectoral and pelvic fins; the distinct, large, irregular spots distinguish it from other morays in the eastern Pacific. *Size*: Length to about 4.3 ft. (140 cm). *Habitat*: Shallow reefs, usually in crevices. *Range*: Tropical Pacific, Islas Revillagigedo, and lower Gulf of California to at least Panama.

FAMILY OPHICHTHIDAE
Snake Eels

22. **TIGER SNAKE EEL**
(tieso manchado) ***Myrichthys tigrinus***
Identification: Presence of small pectoral fins, tubed nostrils, lack of rays in tip of tail, and one to three rows of large, black spots on body separate this eel from moray eels (Family Muraenidae) and other snake eels. *Size*: Length to 1.5 ft. (0.5 m). *Habitat*: In and on shallow, sandy bottoms. *Range*: Throughout tropical Pacific, Indian Ocean, and Red Sea, on this coast from the Gulf of California to Peru and offshore islands. *Natural History*: These eels feed primarily at night on a variety of clams, shrimps and crabs.

23. **PACIFIC SNAKE EEL**
(tieso del Pacifico)
Ophichthus triserialis
Identification: A large mouth and variably-sized dark spots on body separate this snake eel from the other snake eels in the Gulf. *Size:* Length to 3.7 ft. (1.1 m). *Habitat*: Shallow, sandy bottoms. *Range*: Klamath River, California, to Peru.

FAMILY CONGRIDAE
Conger Eels

24. **CORTEZ GARDEN EEL**
(anguila jardin) ***Heteroconger digueti***

Identification: Differs from morays in having a small mouth, pectoral fins, and conspicuous lateral line pores. *Size*: Length to 2.1 ft. (0.6 m). *Habitat*: Sand areas near reefs to about 700 ft (275 m). *Range*: Islas San Benito, Baja California, to central and lower Gulf of California and Panama.

T. Uno

FAMILY ELOPIDAE
Machetes

25. **MACHETE**
Elops affinis

Identification: The single dorsal fin, deeply forked tail, silvery body, and the large mouth with the rear of upper jaw extending beyond rear of eye, are all good field characters. *Size*: Length to 3 ft. (90 cm). *Habitat*: In nearshore surface waters around reefs and over sand bottoms. *Range*: Mandaly Beach, California, to Gulf of California and Peru.

A. Haseltine

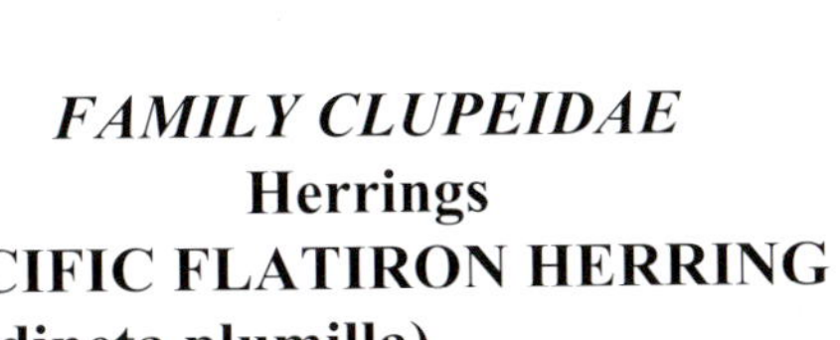

FAMILY CLUPEIDAE
Herrings

26. **PACIFIC FLATIRON HERRING**
(sardineta plumilla)
Harengula thrissina

Identification: The large eyes, deep body, and black spot behind the gill cover are very distinctive. *Size*: Length to 7.3 inches (18 cm). *Habitat*: Nearshore surface waters, in large schools, near reefs, and over soft bottoms. *Range*: San Pedro, California, to Gulf of California and Peru.

T. Uno

FAMILY SYNODONTIDAE
Lizardfishes

27. BANDED LIZARDFISH
(pez lagarto) ***Synodus lacertinus***

Identification: Lizardfishes can be identified by their long, lizardlike bodies and their habit of resting on their pelvic fins. There are about five red-brown bands or blotches on sides above midbody and ten to twelve red or brown blotches on lower body. *Size*: Length to about 1 ft. (0.3 m). *Habitat*: Usually found on sandy bottoms in depths of 70 ft. (27 m). *Range*: Islas San Benito and Gulf of California to Islas Galapagos.

FAMILY ANTENNARIIDAE
Frogfishes

28. BLOODY FROGFISH
(ranisapo sanguineo)

Antennarius sanguineus

Identification: This frogfish usually has a large black spot beneath the rear of the dorsal fin. The rough jaw frogfish *A. avalonis* has a ringed ocellus in the same location (NI). Color varies from rich brown-red to orange to a light cream, and there are large brown spots on the belly. The roughjaw frogfish lacks these spots. *Size*: Length to about 4 inches (10 cm). *Habitat*: Reefs and cobble bottoms. *Range*: Central Gulf of California to Peru and the Islas Galapagos. The less common rough jaw frogfish ranges from Santa Catalina Island, California, to Peru. *Natural History*: Frogfishes lure smaller fishes with a small wormlike "lure" (esca) located just above the middle of the upper jaw next to the first dorsal spine.

FAMILY BELONIDAE
Needlefishes

29. CALIFORNIA NEEDLEFISH
(Agujon) ***Strongylura exilis***

Identification: The very long, knife-like snout with strong teeth on the jaws, very slender elongate body, and single dorsal and anal fins placed far back near tail fin are distinctive. The California halfbeak, *Hyporhampus rosae*, which at a distance looks like a needlefish, has a very short upper jaw. *Size*: Length to 3 ft. (91 cm). *Habitat*: Surface waters, usually only an inch or two beneath surface. *Range*: San Francisco, California, and Gulf of California to Peru.

FAMILY HOLOCENTRIDAE
Squirrelfishes

30. BIGSCALE SOLDIERFISH (candil ojo manchado)

Myripristis berndti

Identification: This distinctive soldierfish has white edges on the soft dorsal, caudal, pelvic and anal fins, and the rear edge of the gill cover is black. *Size*: Length to about 12 inches (30 cm). *Habitat*: In or near crevices and caves during day, over entire reef while foraging at night. *Range*: Tropical Indo-Pacific, on this coast from Islas Revillagigedo to Islas Galapagos. Also reported from the Gulf of California. *Natural History*: The members of this family possess a venomous spine on the lower corner of the cheek. Soldierfish feed on large planktonic animals such as crab larvae.

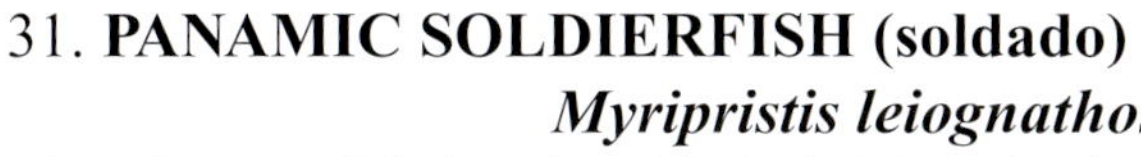

31. PANAMIC SOLDIERFISH (soldado)

Myripristis leiognathos

Identification: Distinguished by its bright red color, large eyes and 2.5 scale rows between lateral line and base of dorsal fin. *Size*: Length to 7 inches (17.8 cm). *Habitat*: During the day soldierfish can be found in caves and crevices; at night they forage around reefs. *Range*: Islas San Benito and Gulf of California to Ecuador.

32. CLARION SOLDIERFISH (candil amarillo)

Myripristis clarionensis

Identification: The Clarion soldierfish can be distinguished from the Panamic soldierfish by counting the scale rows between the lateral line and the base of the dorsal fin. There are 4 rows on the Clarion soldierfish and 3 on the Panamic soldierfish. *Size*: To about 7 inches (18 cm). *Habitat*: Reef crevices and caves during daylight, roams over reef at night. *Range*: Known only from the Islas Revillagigedo and Clipperton Island.

33. **TINSEL SQUIRRELFISH (candil sol)**
Sargocentron suborbitalis

Identification: The prominent spine on the preopercle separates this fish from the Panamic soldierfish; also the silvery body. *Size*: Length to 10 inches (25.4 cm). *Habitat*: Same as Panamic soldierfish. *Range*: Central Gulf of California to Ecuador, and offshore islands.

FAMILY AULOSTOMIDAE
Trumpetfishes

34. **TRUMPETFISH (trumpeteno chino)**
Aulostomus chinensis

Identification: Similar to cornetfish, but this predator on small fish possesses a barbel on the chin and lacks the long caudal fin filament. *Size*: Length to about 32 inches (80 cm). *Habitat*: Occurs around coral and rocky reefs. *Range*: Tropical western and eastern Pacific; on this coast from Islas Revillegigedo to Panama and Islas Galapagos. *Natural History*: These predators often use large fish as "stalking horses" to approach their prey of small fishes and crustaceans.

FAMILY FISTULARIIDAE
Cornetfishes

35. **REEF CORNETFISH (pez corneta)**
Fistularia commersonii

Identification: The shape of the body and the long filament protruding from the caudal fin are very distinctive. *Size*: Length to about 5 ft. (1.6 m). *Habitat*: Around reefs and wrecks to depths of about 100 ft. (30 m). *Range*: Western Indian Ocean to eastern tropical Pacific; on this coast from Bahia Magdalena and Gulf of California, Mexico, to Panama and Islas Galapagos, including all offshore islands. *Natural History*: These stalkers prey on small fishes.

FAMILY SYNGNATHIDAE
Seahorses
36. PACIFIC SEAHORSE
(caballo de mar) ***Hippocampus ingens***

Identification: The best characters are the "horse shaped" head and the body and tail covered with bony rings. The only seahorse on this coast. *Size*: Length to about 9.75 inches (24.7 cm). *Habitat*: Around shallow reefs and other structures, particularly areas where seahorses can hang on to corals, gorgonians, or sponges. *Range*: San Diego, California, to Gulf of California and Peru. *Natural History*: The males carry the eggs in a brood pouch located below the abdomen.

FAMILY SCORPAENIDAE
Scorpionfishes
37. DARKBLOTCH SCORPIONFISH
(rascaclo jugador) ***Scorpaena histrio***

Identification: The most distinguishing feature of this scorpionfish is a dark brown to blackish blotch above the middle of the pectoral fin on the lateral line. The depth of the head exceeds its width and there is a pit between the eyes. *Size*: Length to about 9.5 inches (24 cm). *Habitat*: In or around crevices or on ledges. *Range*: Central Gulf of California to Ecuador and Islas Galapagos. *Natural History*: Scorpionfish usually are night predators, ambushing their prey of small fishes and crustaceans.

38. STONE SCORPIONFISH (lapon)
Scorpaena mystes

Identification: The head is covered with numerous skin flaps or large cirri, which are lacking in the two other shallow water scorpionfishes that occur along the Baja Peninsula, the spotted scorpionfish, *Scorpaena guttata* (NI), and rainbow scorpionfish, *Scorpaenodes xyris*. *Size*: Length to about 18 inches (46 cm). *Habitat*: Shallow reefs and sandy areas. *Range*: Santa Monica Bay, California, and Gulf of California to Ecuador, including offshore islands. *Natural History*: These scorpionfish are ambush predators, darting off the bottom to feed on small fishes.

39. RAINBOW SCORPIONFISH (rascacio arco iris)

Scorpaenodes xyris

Identification: The large, dark spot on the lower rear edge of gill cover and 13 spines in dorsal fin separate this small scorpionfish from the more common stone scorpionfish. *Size*: Length to about 6 inches (46 cm). *Habitat*: Usually in caves and crevices in depths of at least 90 ft. (27 m) during daytime, out in open at night. *Range*: Santa Monica Bay, California, to Gulf of California and Peru, including offshore islands.

FAMILY SERRANIDAE

Sea Basses

40. PACIFIC MUTTON HAMLET (guaseta) *Alphestes immaculatus*

Identification: The best field characters for this secretive and solitary fish are the marked color patterns of the red-brown body with light bars on the pectoral fins. The other hamlet that occurs in the Gulf, the rivulated mutton hamlet, *Alphestes multiguttatus* (NI), has only 5 or 6 bars on the pectoral fins. *Size*: Length to 10 inches (25 cm). *Habitat*: Shallow reefs. *Range*: Islas San Benito and the upper Gulf of California to Peru.

41. PACIFIC SANDPERCH (serrano ucho) *Diplectrum pacificum*

Identification: The dark stripe that runs from the eye to the caudal fin, and a dark blotch at the base of the caudal, are distinctive on juveniles. The stripe in adults tends to fade and be replaced by greyish bars. *Size*: Length to about 10 inches (26 cm). *Habitat*: Usually over sand bottoms, but also at base of reefs in shallow waters to depths of at least 270 ft. (90 m). *Range*: Bahia Magdelena and Gulf of California to Peru. *Natural History*: The members of this family are hermaphroditic--that is, they function as females during their early years and then some individuals transform into males.

42. **LEATHER BASS (mero coriaceo)** ***Dermatolepis dermatolepis***

Identification: In adults, color pattern of white blotches and small dark spots on a gray background (occasionally with several dark bars) is very distinctive. Juveniles have dark bars on a white background. *Size*: Length to about 3 ft. (0.9 m). *Habitat*: Around reefs to depths of at least 100 ft. (30 m). *Range*: Santa Monica Bay, California, to lower Gulf of California, south to Ecuador, and most offshore islands. *Natural History*: Juveniles sometimes shelter in sea urchin spines. These groupers occasionally use grazing fishes to flush out small fishes.

43. **SPOTTED CABRILLA (mero moteado)** ***Epinephelus analogus***

Identification: The large brown spots on the reddish brown body and five indistinct dusky bars on the side are characteristic. *Size*: Length to about 34 inches (87 cm). *Habitat*: Rocky reefs. *Range*: San Pedro, California, to Gulf of California and Peru, including Islas Galapagos. *Natural History*: These solitary groupers feed on fish, crabs and lobsters.

A. Studley

44. **GOLIATH GROUPER (mero guasa)** ***Epinephelus itajara***

Identification: The very small eyes, rounded anal and tail fins, and enormous size of adults are distinctive. The giant sea bass, *Stereolepis gigas* (NI) also reaches great size, but differs from the jewfish by having square anal and tail fins and large eyes. *Size*: Length to about 5 ft. (1.5 m). *Habitat*: In or around caves, in reefs, or around shipwrecks, in depths of 15 to 100 ft. (5 to 30 m). *Range*: Tropical Atlantic and eastern Pacific oceans; on this coast from central Gulf of California to Peru.

45. **FLAG CABRILLA (cabrilla piedrera)**

Epinephelus labriformis

Identification: White spots on the olive-green to red-brown body and red margins on the soft dorsal, anal, and tail fins are very distinctive. *Size*: Length to 1.7 ft. (0.5 m). *Habitat*: Rocky areas in depths to 100 ft. (30 m). *Range*: Rocas Alijos, Baja California, and northern Gulf of California to Peru, including offshore islands.

46. **PANAMA GRAYSBY (enjambre)**

Cephalopholis panamensis

Identification: Distinguished by the 9 to 10 dark bars on the body and blue and orange spots on the side of the head. *Size*: Length to at least 1 ft. (0.3 m). *Habitat*: Rocky reefs and crevices to depths of 250 ft. (76 m). *Range*: Gulf of California to Ecuador, and most offshore islands.

47. **RAINBOW BASSLET (merito arco iris)**

Liopropoma fasciatum

Identification: This small bass has two yellow stripes along the middle of the body separated by a brownish stripe along the lateral line. There is another yellow stripe along the base of the dorsal fin. *Size*: Length to about 6 inches (15 cm). *Habitat*: In and around small reef cracks and crevices. In depths of about 70 to 810 ft. (22 to 250 m). *Range*: Islas San Benito and Gulf of California to Panama.

48. **GULF GROUPER (baya)**

Mycteroperca jordani

Identification: This large grouper is sometimes confused with the broomtail grouper (#50), particularly the juveniles. Gulf groupers lack the notched appearance of the rear edge of the tail fin, due to the protruding fin rays, characteristic of the broomtail grouper. The adults are grey-brown; juveniles have dark blotches on the body. *Size*: Length to about 5 ft. (150 cm). Weight to about 200 lbs (90 kg). *Habitat*: Around rocky reefs in depths of 100 ft. (30 m) or more during the summer, shallow reefs, 15 to 50 ft. (4.6 to 9 m) during the rest of the year. *Range*: La Jolla, California, to Gulf of California and Mazatlan. *Natural History*: Sport and commercial overfishing has reduced the population of these fish predators.

49. **SAWTAIL GROUPER (garropa aserrada)**

Mycteroperca prionura

Identification: Jagged tail fin and large and small red-brown spots on the body are distinctive; can be separated from the less common broomtail grouper, *M. xenarcha* (# 50), by the color pattern: the broomtail grouper is brown to grayish-green with irregular dark blotches. *Size*: Length to about 3 ft. (100 m). *Habitat*: Rocky reefs and caves to depths of at least 160 ft. (50 m). *Range*: Rocas Alijos, Baja California, and Gulf of California.

50. **BROOMTAIL GROUPER (garropa jasplada, mero brujo)**

Mycteroperca xenarcha

Identification: The protruding tail fin rays, which give the rear end of the tail fin a notched appearance, are very distinctive. When these groupers are over the reef, the dark blotches on the body are prominent; when over sand bottoms, the entire body color changes to a light grey or tan and the blotches become indistinct. *Size*: Length to about 40 inches (100 cm). *Habitat*: In and around rocky reefs and mangroves from shallow depths to about 160 ft. (50 m). *Range*: San Francisco Bay to Gulf of California and Peru.

Golden Color Phase

51. LEOPARD GROUPER AND GOLDEN GROUPER (cabrilla sendinera) *Mycteroperca rosacea*

Identification: Small reddish spots cover the body, forming a reticulated pattern; dorsal, anal, and pelvic fins have white margins. A few of the leopard groupers turn yellow gold (golden groupers), and these occasionally have small patches of dark pigment. *Size*: Length to 3 ft. (0.9 m). *Habitat*: Rocky reefs to depths of about 150 ft. (46 m). *Range*: Bahia Magdalena and Gulf of California to Puerto Vallarta.

adult

juvenile

52. PACIFIC CREOLEFISH (rabirubia de lo alto) *Paranthias colonus*

Identification: Salmon-red color, deeply forked tail, and the four or five light spots on the back identify the adult. Juveniles are pinkish-yellow; the dorsal spots are bright blue. *Size*: Length to at least 14 inches (35.6 cm). *Habitat*: Around reefs, usually in schools; during daytime these foraging schools range throughout the water column. In depths to about 200 ft. (61 m). *Range*: Commonly found from Rocas Alijos, off Baja California, and central Gulf of California to Peru and offshore islands. *Natural History*: Creolefish feed on plankton and small fishes. They in turn are fed on by larger fishes and sea birds.

53. **GOLD-SPOTTED SANDBASS (extranjero)** ***Paralabrax auroguttatus***

Identification: The dark brown stripes along the sides and large golden spots identify this small sandbass. The stripes change into rows of large spots in adults. The elongated third dorsal is about 3 times longer than the second spine. *Size*: Length to about 14 inches (35 cm). *Habitat*: Around rocky reefs and sand rock interfaces, in depths of about 50 to 260 ft. (15 to 80 m). *Range*: Southern California to Gulf of California.

54. **SPOTTED SANDBASS (cabrilla de roca)** ***Paralabrax maculatofasciatus***

Identification: This sandbass has dark bars on the body, and the body and head are covered with large black spots. *Size*: Length to 22 inches (56 cm). *Habitat*: Usually on sandy bottoms at base of reefs and shallow depths to 200 ft. 61 m). *Range*: Monterey, California, to Mazatlan, Mexico, and the Gulf of California. Common in Bahia Magdalena and upper Gulf. *Natural History*: This daytime predator feeds on small fishes and crustaceans. Like all members of the family, spotted sandbass reverse sexes, functioning first as females, then transforming into males.

55. **BARRED SERRANO (guaseta serrano)** ***Serranus psittacinus***

Identification: The best field characters are the six pairs of dark bars, often condensed into a dark stripe above lateral line, and eight dark bars below the line. *Size*: Length to about 7 inches (17.8 cm). *Habitat*: Around reefs, on sandy bottoms, to depths of at least 200 ft. (61 m). *Range*: Bahia Magdalena and Gulf of California to Peru, Islas Galapagos, and other offshore islands. *Natural History*: These predators feed on other fishes. Barred serranos are capable of functioning as either males or females.

56. **SOCORRO SERRANO**

Serranus socorroensis

Identification: This handsome small serrano has red bars and a broad red stripe along the midline over a background color of creamy white. *Size*: Length to about 3 inches (8 cm). *Habitat*: Cobble or rubble sand bottoms at base of reefs. In depths of about 70 to 100 ft. (22 to 30 m). *Range*: Known only from Socorro, Islas Revillagigedo.

FAMILY GRAMMISTIDAE
Soapfish

57. **CORTEZ SOAPFISH (jabonero de Cortes)** ***Rypticus bicolor***

Identification: Best distinguished underwater by the rounded dorsal, anal, and caudal fins and brown body with light spots and blotches. The blackfin soapfish, *R. nigripinnis* (NI), is distinguishable by its more regular pattern, and it usually has only two spines in the dorsal fin, while the Cortez soapfish usually has three spines. *Size*: Length to about 8 inches (19.3 cm). *Habitat*: In caves and crevices during the daytime, soapfish come out at night to feed on small fish around reefs and over sand bottoms. Depths to about 225 ft. (69 m). *Range*: Gulf of California to Peru. *Natural History*: Predators tend to avoid soapfishes because of their ability to secrete a toxin.

FAMILY PRIACANTHIDAE
Bigeyes

58. **GLASSEYE (catalufa de roca)**

Heteropriacanthus cruentatus

Identification: The large eye and reddish body and the dark specks in the fin membranes of the dorsal and anal fins distinguish this nocturnal fish from other cave dwellers. *Size*: Length to 1.1 ft. (0.3 m). *Habitat*: Caves and crevices. *Range*: Tropical western Atlantic and eastern Pacific; on this coast southern California to Gulf of California and Islas Revillagigedo south to Ecuador. *Natural History*: These nocturnal predators feed on crustaceans, octopuses, worms and fishes.

59. **POPEYE CATALUFA**

(catalufa semaforo) ***Pristigenys serrula***

Identification: The best field characters are the very large eyes, deep oval, crimson-colored body, and very large pelvic fins. *Size*: Length to 13 inches (33 cm). *Habitat*: Around reefs to depths of 198 ft. (60 m). *Range*: Monterey Bay, California, to Gulf of California and Peru. *Natural History*: Popeye catalufa feed at night on small fishes and crustaceans.

FAMILY APOGONIDAE
Cardinalfishes

60. **PLAIN CARDINALFISH**

Apogon atricaudus

Identification: Similar to Guadalupe cardinalfish but possess a dark blotch on first dorsal fin. *Size*: Length to about 3 inches (7.6 cm). *Habitat*: By day, in crevices, caves, and among sea urchins; found to depths of about 100 ft. (30 m). Feed in open at night. *Range*: Cabo San Lucas and Islas Revillagigedo.

61. **GUADALUPE CARDINALFISH**

Apogon guadalupensis

Identification: Lacks both bar and spot. *Size*: Length to 5 inches (12.7 cm). *Habitat*: In caves and crevices to depths of 60 ft. (18.3 m). *Range*: Farnsworth Bank, California, to Cabo San Lucas.

62. **PINK CARDINALFISH**

Apogon pacifici

Identification: The dark bar below the second dorsal fin does not extend into dorsal fin of this nocturnal fish; lacks spot on caudal peduncle. *Size*: Length to about 4 inches (10 cm). *Habitat*: Caves and crevices during daylight, from shallow waters to about 100 ft. (30 m). *Range*: Islas San Benito to Gulf of California; from Loreto to Cabo San Lucas in Gulf; recorded as far south as Peru.

63. **BARSPOT CARDINALFISH (cardenal)**

Apogon retrosella

Identification: Barspot cardinalfish possess both a bar, located below the second dorsal fin, and a tail spot. The bar extends upward into the dorsal fin. *Size*: Length to about 4 inches (10 cm). *Habitat*: This uncommon cardinalfish can be observed in the same habitat as the pink cardinalfish to depths of 200 ft. (61 m). *Range*: Islas San Benito and Gulf of California, to Mazatlan and Cabo San Lucas; more common in the lower Gulf.

FAMILY MALACANTHIDAE

Tilefishes

64. **FLAGTAIL BLANQUILLO**

Malacanthus brevirostris

Identification: The long slender body and 2 black stripes on the tail fin readily identify this tilefish. *Size*: Length to about 12 inches (30 cm). *Habitat*: Over sand and rubble bottoms at base of reefs. *Range*: Tropical Indo-Pacific. On this coast, Costa Rica to Colombia. *Natural History*: These tilefish construct their own burrows under rocks. Tilefishes feed on a variety of small fishes and invertebrates.

FAMILY ECHENEIDAE
Remoras

65. **REMORA** ***Remora remora***

Identification: This remora is commonly found on mantas. They have a short disc (sucker). The color varies with the background of the host, i.e., when on white underside of mantas, they are creamy-white; when on the darker dorsal side, they are a dark-brown color. *Size*: Length to about 34 inches (86 cm). *Habitat*: Open ocean, usually on a host, such as mantas, sharks, sea turtles or sometimes ships. *Range*: Worldwide in tropical oceans. On this coast from San Francisco to Chile. *Natural History*: This remora feeds on parasitic copepods removed from the host's body, as well as on other small planktonic animals encountered by the host. The remora in turn is cleaned by other fishes when around reefs, such as the Clarion angelfish (#111).

FAMILY CARANGIDAE
Jacks

66. **ISLAND JACK (jurel isleno)** ***Caranx orthogrammus***

Identification: The gold spots above and below the lateral line, but located below the center of soft dorsal fin, is the best underwater character. *Size*: Length to about 2 ft. (0.6 cm). *Habitat*: Found around reefs. *Range*: Tropical Indo-Pacific, Hawaii, and Islas Revillagigedo to Islas Galapagos. *Natural History*: These jacks feed on sand dwelling crabs.

67. **GREEN JACK (cocinero dorado)** ***Caranx caballus***

Identification: Differs from other *Caranx*-type jacks by its longer, more slender body and long pectoral fin. *Size*: Length to 1.3 ft. (0.4 m). *Habitat*: Commonly found around reefs. *Range*: Monterey Bay, California, to Gulf of California and Peru, including offshore islands.

68. **BLACK JACK (jurel negro)**

Caranx lugubris

Identification: The brown to black color is the best underwater character for this nearshore, pelagic jack. *Size*: Length to at least 3 ft. (0.9 m). *Habitat*: Frequently found around reefs and islands. *Range*: Worldwide in tropical waters; on this coast from Rocas Alijos, Baja California, to Islas Galapagos. *Natural History*: Brown jacks are often observed around sewage outfalls of ships; they have also been observed feeding on feces discharged by mantas.

69. **BLUE-SPOTTED JACK (jurel de aleta azul)**

Caranx melampygus

Identification: The blue and black spots on a silvery body are very distinctive on this pelagic jack. *Size*: Length to at least 3 ft. (0.9 m). *Habitat*: Found around reefs. *Range*: Tropical Indo-Pacific; in eastern Pacific from Cabo San Lucas and Islas Revillagigedo to Panama, and Islas Galapagos. *Natural History*: This jack often uses foraging reef fishes to flush out small fishes.

70. **BIGEYE JACK**

Caranx sexfasciatus

Identification: The slender body and curved head with relatively large eyes, completely scaled breast, white tipped dorsal fin, and black spot on rear upper edge of gill cover separate this common jack from the Pacific crevalle jack, *C. caninus* (NI), which has a deeper body, lacks scales on breast except for a small patch, and has a dark blotch at the base of the pectoral fins. *Size*: Length to 30 inches (76 cm). *Habitat*: Occur in schools in mid-depths around reefs. *Range*: Indian and Pacific oceans; on this coast from San Diego, California, and Gulf of California to Ecuador, and offshore islands. *Natural History*: The males turn black during mating season.

71. **RAINBOW RUNNER (macarela salmon)**

Elagatis bipinnulata

Identification: The very slender, elongate body with yellow and blue stripes and the single finlet in back of dorsal and anal fins separate this fish from other members of the jack family. *Size*: Length to about 4 ft. (120 cm). *Habitat*: Surface waters offshore and around islands. *Range*: Tropical seas worldwide; on this coast from Isla Guadalupe, Baja California, to Islas Galapagos.

D. Nelson

72. **GOLDEN JACK (jurel dorado)**

Gnathanodon speciosus

Identification: The juveniles and young adults are silver to yellow, with 7 to 11 dark bars on the sides. *Size*: Length to about 44 inches (110 cm). *Habitat*: Around reefs near the soft bottom and up in water columns. *Range*: Tropical Indo-Pacific; on this coast from the Gulf of California to Ecuador. *Natural History*: These colorful jacks are bottom feeders, rooting in sand or mud for small fishes, crustaceans and molluscs. Young adults often are found "piloting" whale sharks and other large fishes.

T. Hobson

73. **MEXICAN LOOKDOWN (jorobado antena)**

Selene brevoorti

Identification: The very deep, silvery body and deeply sloping forehead are highly distinctive. *Size*: Length to 12 inches (30 cm). *Habitat*: Around reefs and in open water. *Range*: San Diego, California, and lower Gulf of California to Peru.

74. **ALMACO JACK (pez fuerte)**

Seriola rivoliana

Identification: This pelagic jack can be distinguished underwater from the yellowtail, *Seriola lalandi* (NI), by the lack of a yellow tail and the dark bar that runs through the eye. *Size*: Length to about 40 inches (103 cm). *Habitat*: Frequently found around reefs. *Range*: Oceanside, California, to and in the Gulf of California to Peru, including Islas Revillagigedo and Islas Galapagos.

K. McDonnell

75. **YELLOWTAIL (jurel de aleta amarilla)** ***Seriola lalandi***

Identification: This very popular sportfish has yellowish fins, a yellow dusky stripe at midbody and a darker stripe on the head that extends through the eye. *Size*: Length to 5 ft. (1.5 m), weight to 80 lbs. (36 kg). *Habitat*: Usually near surface around reefs and kelp beds, to depths of 228 ft. (69 m). *Range*: Worldwide in tropical waters; on this coast from British Columbia to Gulf of California and Chile. *Natural History*: Yellowtail feed on a variety of fishes, squid, pelagic red crabs and other invertebrates.

76. **GAFFTOPSAIL POMPANO (pompanito)** ***Trachinotus rhodopus***

Identification: The only jack having soft dorsal and anal fin rays that reach beyond the base of tail fin if laid flat. *Size*: Length to 2 ft. (60 cm). *Habitat*: Commonly found around reefs. *Range*: Zuma Beach, California, to Gulf of California and Peru.

77. **STEEL POMPANO**

(pampano acerado) ***Trachinotus stilbe***

Identification: This small jack has a distinctive white bar that runs from the back to just behind the gill cover. *Size*: Length to about 16 inches (40 cm). *Habitat*: Pelagic, usually only a few feet below surface, around offshore islands. *Range*: Islas Revillagigedo to Islas Galapagos and Ecuador.

FAMILY NEMATISTIIDAE
Roosterfish

78. **ROOSTER FISH (papa gallo)**

Nematistius pectoralis

Identification: The very long filamentous spines in the dorsal fin and two curved dark stripes on the body that originate at the base of the spinous dorsal fin and descend to about midbody, then run posteriorly are distinctive. *Size*: Length to about 4 ft. (122 cm). *Habitat*: Shallow waters, usually over sand, occasionally around reefs. *Range*: San Clemente, California, to Gulf of California and Peru. *Natural History*: One of the most popular game fishes in the Sea of Cortez because of its fighting ability. Feeds on small to medium-size fishes such as herring.

FAMILY LUTJANIDAE
Snappers

79. **BARRED PARGO (pargo coconaco)**

Hoplopagrus guntheri

Identification: The upper body is green-brown to dark brown; the lower body ranges from copper-red to maroon. There are eight or nine bars on the body. Juveniles have a dark spot below the rear of the soft dorsal fin. *Size*: Length to 2.5 ft. (0.8 m). *Habitat*: Juveniles can be observed in mangrove swamps; adults occur around reefs, large boulders, and caves. To depths of about 90 ft. (27 m). *Range*: San Diego, California, to Gulf of California and south to Panama, and Islas Galapagos. *Natural History*: These predators stay in caves and crevices during the day and feed on crustaceans and fishes at night.

80. MULLET SNAPPER (pargo raicero)

Lutjanus aratus

Identification: The more slender body, distinctive stripes on the body, and 11 spines in the dorsal fin separate this fish from the other snappers. *Size*: Length to 30 inches (76 cm). *Habitat*: Juveniles in estuarine waters; adults in schools in midwater around reefs. *Range*: Punta Eugenia, Baja California, and central Gulf of California to Ecuador.

81. YELLOW SNAPPER (pargo amarillo)

Lutjanus argentiventris

Identification: The yellow dorsal, anal, pectoral, and caudal fins, and the yellow rear half of the body are the best field identification characters for adults. *Size*: Length to 2 ft. (0.6 m). *Habitat*: Around reefs and over sand bottoms near reefs, also in caves; to depths of at least 75 ft. (23 m). *Range*: Oceanside, California, and upper Gulf of California to Peru. *Natural History*: These snappers feed during the day on octopuses, shrimps, crabs and fishes.

82. SPOTTED ROSE SNAPPER (pargo lunarejo) ***Lutjanus guttatus***

Identification: The large dark spot below the dorsal fin is very distinctive; however, on some fish this spot may almost disappear. *Size*: Length to 30 inches (76 cm). *Habitat*: Over sand bottoms and near reefs and wrecks. *Range*: Gulf of California to Peru.

83. **GOLDEN SNAPPER**

(pargo barbirebia) ***Lutjanus inermis***

Identification: This snapper usually has a broad golden-yellow stripe along the midline, beginning below the spinous dorsal fin and extending onto the tail fin. *Size*: Length to about 14 inches (35 cm). *Habitat*: Usually in schools or large aggregations over reefs. *Range*: Southern Gulf of California to Panama. *Natural History*: Some juveniles mimic the coloration of the scissortail damselfish (#117). Snappers feed at night, primarily on other fish.

84. **PACIFIC DOG SNAPPER**

(pargo negro) ***Lutjanus novemfasciatus***

Identification: Can best be identified underwater by the large canine teeth, which are usually visible, and the nine obscure, dusky bars on a silvery body. Large adults are reddish when out of water. *Size*: Length to 3 ft. (0.9 m). *Habitat*: In caves during the day; feed around reefs at night to depths of at least 100 ft. (30 m). *Range*: Southern California and Gulf of California to Peru. *Natural History*: These predators forage at night for fishes and crustaceans.

85. **BLUE-AND-GOLD SNAPPER**

(pargo rayado) ***Lutjanus viridis***

Identification: The blue and gold stripes are very distinctive. *Size*: Length to 1 ft. (0.3 cm). *Habitat*: Schools occur around reefs to depths of about 50 ft. (15 m). *Range*: This common snapper occurs from Bahia Tortugas, on the outer coast of Baja California, and from the central Gulf (Isla Monserrate) to Ecuador, and offshore islands.

FAMILY GERREIDAE
Mojarras

86. YELLOWFIN MOJARRA
(mojarra blanca) ***Gerres cinereus***

Identification: Distinguished by the long snout with a protrusible mouth and the dark bars on the sides of the body. *Size*: Length to 12 inches (0.3 m). *Habitat*: Shallow, sandy bottoms in bays and mangrove swamps. *Range*: Southern Baja California and Gulf of California south to Peru.

FAMILY HAEMULIDAE
Grunts

87. SARGO (burro piedrero)
Anisotremus davidsonii

Identification: The single black bar that extends from the base of the dorsal fin to the pectoral fin readily identifies this schooling grunt. *Size*: Length to 23 inches (58 cm). *Habitat*: In schools over and around reefs, kelp beds and other structures such as piers and jetties, to depths of 130 ft. (40 m). *Range*: Santa Cruz, California, to Gulf of California. *Natural History*: These nocturnal feeders prey on small bottom invertebrates.

88. BURRITO GRUNT (burro frijol)
Anisotremus interruptus

Identification: The sharp, slanting profile of the head, the fleshy lips, and large conspicuous scales with dark spots on their anterior margin readily identify this grunt. *Size*: Length to 1.5 ft. (0.5 m). *Habitat*: The adults are found around reefs and in caves to depths of about 75 ft. (23 m). *Range*: Bahia Sebastian Vizcaino and Gulf of California to Peru; common around Islas Revillagigedo. *Natural History*: This grunt preys on invertebrates and occasionally small fishes at night.

89. **PANAMIC PORKFISH (burro bandera)** ***Anisotremus taeniatus***

Identification: The distinctive color pattern of gold and blue stripes on the body readily identifies this fish. *Size*: Length to about 1 ft. (0.3 m). *Habitat*: Around reefs to depths of at least 75 ft. (23 m). *Range*: Bahía Magdalena, Baja California, and Cabo San Lucas area (Bahia Chileno) to Ecuador.

juvenile adult

90. **CORTEZ GRUNT (burro de Cortes)** ***Haemulon flaviguttatum***

Identification: The pearly-blue spot on each scale is the best character. *Size*: Length to about 16 inches (41 cm). *Habitat*: Around reefs to depths of about 50 ft. (15 m); and, commonly, in mangrove swamps. *Range:* San Diego, California, and Gulf of California to Panama.

91. **SPOTTAIL GRUNT (ronco soldadito)** ***Haemulon maculicauda***

Identification: Distinguished by spots on scales that form stripes on sides, and the dusky blotch at the base of the tail fin. *Size*: Length to about 1 ft. (0.3 m). *Habitat*: During the day, schools occur around reefs; at night these fish move off the reef to feed over sandy bottoms, in depths to at least 100 ft. (30 m). *Range*: Isla Cedros and Gulf of California to Ecuador.

adult

92. **GRAYBAR GRUNT (burro almejera)**

Haemulon sexfasciatum

Identification: The distinctive yellow and dark gray bars are the best field characters. *Size*: Length to 1.5 ft. (0.5 m). *Habitat*: Around reefs, over sandy bottoms to depths of abut 75 ft. (23 m). *Range*: Bahia Magdalena and Gulf of California to Panama.

juvenile

93. **LATIN GRUNT (ronco chere-chere)**

Haemulon steindachneri

Identification: This grunt differs from the lookalike Cortez grunt (# 90) by having narrow brown stripes in oblique rows corresponding with the scale rows. The caudal fin is yellow with a black spot. *Size*: Length to about 12 inches (30 cm). *Habitat*: Usually in schools around reefs or over sand at base of reefs. Shallow depths. *Range*: Baja California to Peru.

94. **WAVYLINE GRUNT (ronco jopaton)**

Microlepidotus inornatus

Identification: This grunt has 7 to 9 orange stripes on the sides of the body; some of the stripes above the lateral line form a wavy pattern. *Size*: Length to about 12 inches (30.5 cm). *Habitat*: Over sand in loose feeding aggregations at night, around reefs during the day. *Range*: Bahia Magdalena to at least Manzanillo, Mexico. *Natural History*: Wavy line grunts feed on a variety of invertebrates, including planktonic crustaceans and molluscs on the bottom.

95. **SALEMA (ojoton)**

Xenistius californiensis

Identification: This salema has 6 to 8 dark orange stripes along its sides. *Size*: Length to 10 inches (25 cm). *Habitat*: Around shallow, rocky and sandy bottoms in 4 to 35 ft. (1 to 11 m). *Range*: Monterey Bay, California, to Gulf of California and Peru. *Natural History*: These small grunts are midwater feeders, often mixing in with schools of other grunts.

C. Turner

FAMILY SPARIDAE
Porgies

96. **PACIFIC PORGY (mojarron)**

Calamus brachysomus

Identification: Distinguished by the sharp, slanting profile of the head and deep body. *Size*: Length to about 15 inches (38.5 cm). *Habitat*: Usually found over sand around reefs to depths of 225 ft. (69 m). *Range*: Oceanside, California, and Gulf of California to Peru, and Islas Galapagos. *Natural History*: Pacific porgies feed on sand dwelling clams, snails and crustaceans.

FAMILY SCIANIDAE
Drums and Croakers

97. **YELLOWEYE CROAKER (bombache ojo amarillo)**

Odontoscion xanthops

Identification: This dark-colored croaker has a large eye, diameter about 1/2 to 2/5 of length of head, and brown stripes on the body. The stripes are oblique on upper body. *Size*: Length to about 12 inches (30 cm). *Habitat*: Rocky and coral reefs. *Range*: Gulf of California to Panama.

juvenile above adult below

98. **ROCK CROAKER (corvinita gungo)**

Pareques viola

Identification: The light to dark brown coloration and high first dorsal fin identify the adults; the yellow-and-black striped juveniles are very distinctive. *Size*: Length to about 10 inches (25 cm). *Habitat*: In caves and crevices to depths of at least 100 ft. (30 m). *Range*: Islas San Benito and Gulf of California to Panama, including Islas Revillagigedo.

FAMILY MULLIDAE
Goatfish

99. **MEXICAN GOATFISH (salmonete barbon)**

Mulloidichthys dentatus

Identification: The yellow stripe running from the eye to the tail readily identifies this goatfish. At night the body is covered with reddish blotches, but the yellow stripe is still visible. *Size*: Length to about 1 ft. (0.3 m). *Habitat*: Usually over sandy bottoms around reefs to depths of at least 75 ft. (23 m). *Range*: Southern California and the Gulf of California to Peru, including the offshore islands. *Natural History*: Goatfishes use their barbels to ferret out small sand and mud dwelling invertebrates and fishes.

FAMILY KYPHOSIDAE
Sea Chubs

100. **GULF OPALEYE (ojo azul)**

Girella simplicidens

Identification: The dark body, bright blue eyes, and the three to four white spots (sometimes absent) on the sides just below the dorsal fin are good field characters for this abundant upper Gulf of California fish. *Size*: Length to about 1.5 ft. (0.5 m). *Habitat*: Shallow reefs in areas where algae are abundant. *Range*: Gulf of California.

101. **ZEBRA PERCH (chopa bonita)** ***Hermosilla azurea***

Identification: The scales on head and the presence of bars on the body separate this fish from all other members of the family. *Size*: Length to 1.4 ft. (0.4 m). *Habitat*: Shallow reefs, particularly in the mid- and upper-Gulf. *Range*: Klamath River, California, to Gulf of California.

102. **BLUE-BRONZE CHUB (chopa gris)** ***Kyphosus analogus***

Identification: Can best be distinguished from the other chub that occur in the Gulf of California by the presence of an orange stripe that runs from the upper jaw, under the eye, to the back edge of the gill cover. The thin, light orange stripes on the sides are also good field characters. *Size*: Length to about 14 inches (35 cm). *Habitat*: Around shallow reefs. *Range*: Redondo Beach, California, to the Gulf of California and Peru.

103. **CORTEZ CHUB (chopa de Cortes)** ***Kyphosus elegans***

Identification: Lacks the orange stripe under the eye and orange body stripes of the blue-bronze chub; instead has a faint dark stripe under the eye. Another good character for separating these two chubs is based on the outer edge of the anal fin: if a straight line is drawn along the outer edge of the anal fin it will pass through the upper part of the tail fin of the blue-bronze chub, but will pass above the tail fin of the Cortez chub. *Size*: Length to 1.3 ft. (0.4 m). *Habitat*: Around shallow reefs. *Range*: Gulf of California to Ecuador and Islas Galapagos.

104. SOCORRO CHUB (chopa de Socorro)

Kyphosus lutescens

Identification: Dark phase is indistinguishable from the Cortez chub, but the golden color phase is unique; about 1 of 10 Socorro chubs are in golden phase. *Size*: Length to about 1.3 ft. (0.4 m). *Habitat*: Shallow reefs to depths of about 100 ft. (30 m). *Range*: Rocas Alijos to Islas Revillagigedo.

105. RAINBOW CHUB (chopa salema) ***Sectator ocyurus***

Identification: The elongated body with blue and yellow horizontal bands, yellowish fins, and deeply forked tail are good field characters. *Size*: Length to about 24 inches (60 cm). *Habitat*: Pelagic; around islands and reefs. *Range*: Redondo Beach, California, south to Cabo San Lucas to Panama and Islas Galapagos.

FAMILY EPHIPPIDAE
Spadefishes

106. PACIFIC SPADEFISH (poguala peluquero, chambo)

Chaetodipterus zonatus

Identification: The dark bars on the sides and the long pelvic fins readily identify this spadefish. *Size*: Length to about 26 inches (65 cm). *Habitat*: Usually over sand or rubble bottoms, in bays and other nearshore areas; occasionally around reefs. *Range*: San Diego, California, to Gulf of California and Peru. *Natural History*: The Pacific spadefish occasionally can be observed around reefs being cleaned of parasites by other fishes, such as juvenile Mexican hogfish (#131).

FAMILY CHAETODONTIDAE
Butterflyfishes

107. **SCYTHE BUTTERFLYFISH (mariposa guadana)**

Prognathodes falcifer

Identification: The high, spinous dorsal fin, long snout, and scythe-shaped black markings on sides of body are very distinctive. *Size*: Length to 6 inches (15.2 cm). *Habitat*: Deep reefs and kelp beds; in depths of 35 to 250 ft. (11 to 76 m). *Range*: Santa Catalina Island, California, to Islas Galapagos; common around Islas San Benito.

108. **THREEBANDED BUTTERFLYFISH (mariposa muneca)**

Chaetodon humeralis

Identification: The three dark bars on a silvery body are very distinctive. *Size*: Length to 10 inches (25.4 cm). *Habitat*: Usually around reefs and wrecks; to depths of about 180 ft. (55 m). *Range*: La Jolla Canyon, California, Bahia Kino in the Gulf of California and south to Peru including Islas Galapagos. *Natural History*: Butterflyfish eggs are pelagic and hatch in about 1 or 2 days. The pelagic larval butterflyfish will settle on the reef in about 40 days.

109. **LONGNOSE BUTTERFLYFISH (mariposa hocicona)**

Forcipiger flavissimus

Identification: Easily recognized by the long, slender snout, yellow body, and black tail. *Size*: Length to 8.5 inches (22 cm). *Habitat*: Rocky reefs to depths of about 75 ft. (23 m). *Range*: Indo-Pacific; on this coast Gulf of California from La Paz to Cabo San Lucas, to Islas Galapagos and other offshore islands.

110. **BARBERFISH (mariposa barbero)**
Johnrandallia nigrirostris

Identification: The black forehead and the silvery-yellow body with the stripe under the dorsal fin are the best field characters. *Size*: Length to 8 inches (20.3 cm). *Habitat*: Shallow rock and coral reefs; to depths of about 130 ft. (40 m). *Range*: Rocas Alijos, Baja California, and central Gulf of California to Panama, including offshore islands. *Natural History*: Barberfish are very active cleaners.

juvenile

FAMILY POMACANTHIDAE
Angelfishes

111. **CLARION ANGELFISH (angel de Clarion)**
Holacanthus clarionensis

Identification: The bright orange body and tail of the adults are very distinctive. Juveniles resemble juvenile king angelfish, but differ in having an orange head and tail. *Size*: Length to 1 ft. (0.3 m). *Habitat*: Rocky and coral reefs to depths of about 100 ft. (30 m). *Range*: Isla Guadalupe on outer coast and from Isla Espiritu Santo in Gulf of California to Islas Revillagigedo and Isla Clipperton. *Natural History*: Clarion angelfish adults occasionally clean parasites from other fishes, including mantas.

adult

112. **KING ANGELFISH (angel real)**

Holacanthus passer

Identification: The adults have a yellow tail and a vertical white bar on the sides just behind the origin of the pectoral fin. Juveniles have more yellow on the body and yellow pelvic fins. *Size*: Length to 12 inches (30 cm). *Habitat*: Around shallow reefs to depths of about 160 ft. (49 m). *Range*: Isla Guadalupe, Baja California, and central Gulf of California to Ecuador and Islas Galapagos. *Natural History*: This grazer feeds on a variety of algae and invertebrates, including planktonic forms.

juvenile

adult

juvenile Cortez angelfish

adult

113. **CORTEZ ANGELFISH (angel de Cortes)**

Pomacanthus zonipectus

Identification: The gray body of adults, with yellow and black bands on the head and anterior body, and the yellow and black bands on the head and body of juveniles are distinctive characters. *Size*: Length to 1.5 ft. (0.5 m). *Habitat*: Around reefs to depths of at least 100 ft. (30 m). *Range*: Long Beach, California, and Gulf of California to Peru. *Natural History*: Adults are grazers, feeding on a variety of algae, bryozoans, hydroids, sponges, and tunicates. The juveniles obtain some of their food by picking parasites off other fishes.

FAMILY POMACENTRIDAE
Damselfishes

114. **PANAMIC SERGEANT MAJOR**
(pintano) ***Abudefduf troschelii***

Identification: The five or six black bars on a silver to yellow body are good field characters. *Size*: Length to about 7 inches (18 cm). *Habitat*: Large aggregations occur around and over reefs and wrecks to depths of about 100 ft. (30 m). *Range*: Redondo Beach, California, and Gulf of California to Peru, including the offshore islands. *Natural History*: Most damselfish make nests by cleaning off an area on the reef for the female to deposit her eggs. The eggs hatch in 2 to 7 days; the larvae are pelagic, drifting near the surface for up to 50 days.

115. **SWALLOWTAIL DAMSELFISH**
Azurina hirundo

Identification: The long, slender, steel-blue body, deeply forked tail fin, and light yellow pectoral fins of this damselfish are good field characters. *Size*: Length to about 7 inches (18 cm). *Habitat*: In midwater, around reefs; occasionally mixed with blacksmith, *Chromis punctipinnis* (NI). *Range*: San Clemente Island and possibly Santa Catalina Island, California, to at least Islas Revillagigedo; not recorded from Gulf of California.

116. **SILVERSTRIPE CHROMIS**
(castaneta oval) ***Chromis alta***

Identification: The iridescent blue juveniles resemble Cortez damselfish juveniles but lack the dark spot beneath the soft dorsal fin. Adults may retain some of the blue markings but also have white lines at the base of the dorsal and anal fins. These lines may sometimes be absent. *Size*: Length to about 6 inches (15 cm). *Habitat*: Found around deeper reefs to depths of about 200 ft. (61 m). *Range*: Isla Guadalupe to Cabo San Lucas; Isla Catalina in Gulf of California to Islas Galapagos.

117. **SCISSORTAIL DAMSELFISH (castaneta conquita)**

Chromis atrilobata

Identification: The best field characters are the deeply forked tail and the white spot below the soft dorsal fin. *Size*: Length to 5 inches (12.7 cm). *Habitat*: These damselfish congregate in large numbers around reefs; found to depths of about 250 ft. (76 m). *Range*: Upper Gulf of California and Islas San Benito on outer coast of Baja California to northern Peru, including Islas Revillagigedo and Islas Galapagos.

118. **BLUE-AND-YELLOW CHROMIS (castaneto de Limbaugh)**

Chromis limbaughi

Identification: The best field characters are the bright blue anterior body and the yellow dorsal fin and posterior upper body. *Size*: Length to about 6 inches (15 cm). *Habitat*: Aggregations occur around reefs to depths of about 225 ft. (69 m). *Range*: Bahia de los Angeles to Cabo San Lucas.

119. **BUMPHEAD DAMSELFISH (jaqueta vistosa)**

Microspathodon bairdi

Identification: Distinguished by blue eyes, the bump on the head, and the lack of filamentous dorsal, caudal, and anal fins. Juveniles resemble young beaubrummel (#122), having bright, iridescent blue above and bright orange below. *Size*: Length to about 1 ft. (0.3 m). *Habitat*: Found around reefs. *Range*: Central Gulf of California south to Ecuador, and offshore islands.

juvenile

adult

120. **GIANT DAMSELFISH (castanuela gigante)**

Microspathodon dorsalis

Identification: Adults are distinguished by the long dorsal, caudal, and anal fin rays; adult males have silvery-white heads. The juveniles are identified by iridescent blue spots on a dark body. *Size*: Length to 1 ft. (0.3 m). *Habitat*: Giant damselfish live around reefs; to depths of about 75 ft. (23 m). *Range*: Bahia Magdalena, Baja California, and central Gulf of California to Ecuador and Islas Galapagos.

121. **ACAPULCO DAMSELFISH (jaqueta Acapulco)**

Stegastes acapulcoensis

Identification: The adults have a dark brownish head and rear body. There is a wide light yellowish band that extends from the rear of the gill cover to about mid-body and white upper edges of the pectoral fins. Juveniles are similar to juveniles of the Cortez damselfish. *Size*: Length about 7 inches (17 cm). *Habitat*: Nearshore rocky reefs. *Range*: Bahia Concepcion, Gulf of California, to Peru including Isla Cocos and Islas Galapagos.

122. **BEAUBRUMMEL (pez de dos colores) (jaqueta de dos colores)**

Stegastes flavilatus

Identification: The adults resemble Cortez damselfish but the fin margins are usually yellow. Juveniles are easily identified by the bright blue upper body and yellow lower body. *Size*: Length to about 5.5 inches (14 cm). *Habitat*: Solitary juveniles and adults occur around reefs to depths of about 125 ft. (38 m). *Range*: Bahia Magdalena on outer coast of Baja California, and Gulf of California (Isla Danzante) to Ecuador.

juvenile

adult

juvenile whitetail damselfish

adult

123. **WHITETAIL DAMSELFISH (jaqueta rabo blanco)**

Stegastes leucorus

Identification: The adults have white to yellow margins on the pectoral fins and a pale band across the caudal peduncle. The juveniles are more iridescent and the greenish back coloration is more brilliant than on the adults. *Size*: Length to about 5 inches (13 cm). *Habitat*: This species is another reef dweller, found to depths of about 50 ft. (15 m). *Range*: Isla Guadalupe off Baja California, and central Gulf of California to Islas Revillagigedo and Mazatlan.

juvenile

124. **CORTEZ DAMSELFISH**
(jaqueta de Cortes)

Stegastes rectifraenum

Identification: The best character for the adults is the brown body; fin margins occasionally retain the iridescent blue of the juveniles, which can be separated from other damselfish juveniles by the black spot at the base of the soft dorsal fin and the spot on the dorsal side of the caudal peduncle. *Size*: Length to about 5 inches (13 cm). *Habitat*: Shallow reefs in the central Gulf to depths of about 75 ft. (23 m). *Range*: Bahia Magdalena, Baja California, to Gulf of California.

adult

juvenile clarion damselfish

adult

125. **CLARION DAMSELFISH**
(jaqueta de Clarion)

Stegastes redemptus

Identification: The juveniles resemble the whitetail damselfish, but in the adult the rear portions of the body and the tail fin are yellowish, and the body scales have dark margins which give them a reticulated appearance. *Size*: Length to about 5 inches (13 cm). *Habitat*: Around shallow reefs. *Range*: Cabo San Lucas (rare) and Islas Revillagigedo.

FAMILY CIRRHITIDAE
Hawkfishes

126. LONGNOSE HAWKFISH (halcon narigon) *Oxycirrhites typus*

Identification: The coloration of this hawkfish is similar to the coral hawkfish, but the long pointed snout distinguishes it from that species. *Size*: Length to about 5 inches (13 cm). *Habitat*: This deep-water hawkfish seems to prefer the gorgonians with yellow polyps that occur in depths of about 90 to 100 ft. (27 to 30 m). *Range*: Tropical Pacific Ocean; on this coast from Thetis Bank off outer Baja California, and Gulf of California to Colombia and Islas Galapagos.

127. GIANT HAWKFISH (chino mero) *Cirrhitus rivulatus*

Identification: The brown bands on a gray-brown body and maze of wavy blue lines on the body and fins are good field characters. *Size*: Length to 1.7 ft. (0.5 m). *Habitat*: Commonly observed on reefs, at the edge of crevices, and at entrance to caves; to depths of about 75 ft. (23 m). *Range*: Rocas Alijos, Baja California, and upper Gulf of California to Islas Galapagos.

128. CORAL HAWKFISH (halcon de coral) *Cirrhitichthys oxycephalus*

Identification: The red spots on a whitish background are good field characters. *Size*: Length to about 3.5 inches (8.5 cm). *Habitat*: Occur on coral heads or around crevices and small rocks; to depths of about 75 ft. (23 m). *Range*: Tropical Indo-Pacific; on this coast from Rocas Alijos, Baja California, and central Gulf of California (Islas Santa Ines) to Islas Galapagos including the offshore islands.

FAMILY MUGILIDAE

Mullets

129. MULLET (lisa, liseta) ***Mugil* sp.**

Identification: There are at least five species of mullets that occur in the area covered by this book. Mullets have small mouths, elongated bodies, two dorsal fins, a forked tail fin, and fatty tissue forming the eyelids. *Size*: Length to 3 ft. (0.9 m). *Habitat*: Over shallow sandy bottoms, sometimes just outside surf lines; to depths of 400 ft. (120 m). *Range*: California to South America.

FAMILY SPHYRAENIDAE

Barracudas

130. MEXICAN BARRACUDA (picuda agujona)

Sphyraena lucasana

Identification: The best field characters are the long, slender body with dark bars and large mouth with many teeth. *Size*: Length to about 28 inches (70 cm). *Habitat*: Occur in large schools in surface waters around reefs, wrecks, and other structures. *Range*: Gulf of California. *Natural History*: Barracuda are predators on other fishes. They are considered very edible, but large individuals can be the source of ciguatera poisoning.

juveniles

FAMILY LABRIDAE

Wrasses

131. MEXICAN HOGISH (vieja de piedra)

Bodianus diplotaenia

Identification: This fish can best be identified by the large fleshy hump on the head of the males and the two dark stripes on the bright yellow body of both juveniles and females. *Size*: Length to 2.5 ft. (0.8 m). *Habitat*: These large wrasses prefer shallow reefs, but they have been recorded to depths of about 250 ft. (76 m). *Range*: Isla Guadalupe, Baja California, and Gulf of California to Chile, including most offshore islands. *Natural History*: Most wrasses function as females when first mature, but after a few years most of these females transform into males. Spawning takes place in midwater.

female Mexican hogfish

male Mexican hogfish

132. **WOUNDED WRASSE (senorita herida)**

Halichoeres chierchiae

Identification: Adult males are best identified by the dark-green spot with a red blotch behind the pectoral fin. The females have yellowish bands on the body and an orange tail. *Size*: Length to 8 inches (20.3 cm). *Habitat*: These fish prefer shallow reefs that support growths of algae and that are surrounded by sand patches; occasionally found to depths of 225 ft. (69 m). *Range*: Puerto Penasco in upper Gulf of California to Panama.

female

male

female above
juvenile above

male below
adult below

133. CHAMELEON WRASSE (senorita camaleon)

Halichoeres dispilus

Identification: The pink body, salmon-colored head and caudal fin, and green or blue spot just above the center of the pectoral fin are very distinctive. *Size*: Length to about 9.5 inches (25 cm). *Habitat*: Prefer shallow reef areas adjacent to sandy bottoms; found to depths of about 250 ft. (76 m). *Range*: Islas San Benito on outer coast of Baja California, and Gulf of California to Peru, including Isla Cocos and Islas Galapagos.

D. Dvorak-Cordell Expeditions female

134. SOCORRO WRASSE (doncella de Socorro)

Halichoeres insularis

Identification: The adults of this colorful wrasse have a prominent yellow stripe that runs from just below the eye along the midline to the base of the tail fin. Juveniles are lighter in coloration and the stripe originating below the eye is silver-white. *Size*: Length to about 3.5 inches (8.2 cm). *Habitat*: Rocky reefs to depths of about 80 ft. (25 m). *Range*: Known only from Islas Revillagigedo.

135. INDIGO WRASSE *Pseudojulis* sp.

Identification: This undescribed wrasse is distinguished by the iridescent indigo blue stripe at the base of the dorsal fin that extends to the tail fin in females. Males have yellow bodies with a black bar under the center of the dorsal fin. *Size*: Length to about 5 inches (13 cm). *Habitat*: Around reefs, to depths of at least 60 ft. (18 cm). *Range*: Isla Guadalupe, Baja California, to the Islas Revillagigedo. This was the most common wrasse observed by divers at Rocas Alijos, off Bahia Magdelena, in 1993.

136. **SPINSTER WRASSE (duncella soltera)**

Halichoeres nicholsi

Identification: The male is darker on the anterior of a bluish-green body and has a yellow blotch anterior to the bar (fish from the Islas Revillagigedo have the yellow blotch on the gill cover); the female has a black bar that meets a black body stripe. Juveniles have a pale-yellow body with dark blotches. *Size*: Length to 1.3 ft. (0.4 m). *Habitat*: Adults prefer reefs associated with sandy bottoms; found to depths of about 175 ft. (53 m). *Range*: Isla Guadalupe on outer coast of Baja California, and Gulf of California to Panama, including the offshore islands.

juvenile

female

male

137. **ROCK WRASSE (cocinero)**

Halichoeres semicinctus

Identification: Juvenile rock wrasse have wide, dark stripes on the sides and a silver-white stripe along the midline. Females have rows of black spots on the upper body and males have a distinctive wide, black band that runs down the side from the lateral line passing under the pectoral fin down the belly. *Size*: Length to 15 inches (38 cm). *Habitat*: Around reefs and over rubble and sand bottom, to depths of 78 ft. (24 m). *Range*: Point Conception, California, to Gulf of California. *Natural History*: The females change into males at about 5 years of age. These wrasse sleep under the sand at night with just the head protruding.

female above male below

Adult Juvenile

138. ROCKMOVER WRASSE (duncella alguera)

Novaculichthys taeniourus

Identification: The juveniles of this species are also called "dragon wrasses" because of the greatly extended rays of the dorsal and anal fins. The adults are readily identified by their light-grey heads with brown bands radiating from the edges and the white band across the base of the tail fin. *Size*: Length to about 12 inches (30 cm). *Habitat*: Around reefs and sand or rubble bottoms. *Range*: Indo-Pacific, on this coast from lower Gulf of California to Panama. *Natural History*: True to its name, this wrasse uses its mouth to move rocks in search of invertebrates and other food items. Like many wrasses, the rockmover wrasse will burrow into the sand to escape predators.

139. CALIFORNIA SHEEPHEAD (vieja de California)

Semicossyphus pulcher

Identification: Juveniles have black blotches in the rear of the dorsal fin, base of tail fin and on the anal fin. Adults are light to dark brick-red in general color, and the males have black heads and the posterior of the body is also black. *Size*: Length to 3 ft. (1 m). *Habitat*: Around reefs, kelp beds and over sand during feeding forays; in shallow waters to depths of 289 ft. (88 m). *Range*: Monterey, California, to Gulf of California. *Natural History*: Sheephead feed on sea urchins, molluscs, lobsters and crabs. They can live as long as 50 years or more.

female below male right

140. **ISLAND WRASSE (vieja crepusculo)**

Thalassoma grammaticum

Identification: The males and females are blue-green to green with dark pink heads. There are dark pink bands on the dorsal fin, base of anal fin and on the upper and lower edges of the tail fin. *Size*: Length to about 9.5 inches (24 cm). *Habitat*: Rock and coral reefs. *Range*: Rocas Alijos, Baja California, and Gulf of California to Panama and Islas Galapagos.

female above male bclow

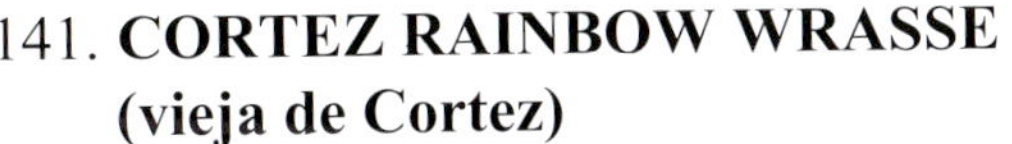

141. **CORTEZ RAINBOW WRASSE (vieja de Cortez)**

Thalassoma lucasanum

Identification: The distinctive yellow and red stripes are the best field characters. Secondary males have purple heads and a yellow band just behind the head. *Size*: Length to 6 inches (15.2 cm). *Habitat*: Shallow reefs to depths of about 160 ft. (40 m). *Range*: Rocas Alijos, on outer coast of Baja California, and central Gulf of California to Islas Galapagos. *Natural History*: Juveniles occasionally obtain food by picking parasites off other fishes.

female above male below

142. **GREEN WRASSE (vieja esmeralda**

Thalassoma virens

Identification: Males are a uniform, bright green color without streaks or spots on head or body and have a deeply lunate tail fin. Females are dull green, the body scales have red streaks and there are 6 faint dark bars on the sides. *Size*: Length to about 11.5 inches (29 cm). *Habitat*: Reefs. *Range*: Known only from Islas Revillagigedo and Isla Clipperton.

juvenile above adult below

143. **PEACOCK WRASSE**
(cuchillo pavo real) ***Xyrichthys pavo***
Identification: The two very long spines of the first dorsal fin and distinctive body shape distinguish this wrasse from the others in the Gulf of California. Adults are light blue or gray with three distinctive bars on sides. *Size*: Length to 10 inches (25.4 cm). *Habitat*: Both adults and juveniles occur over and in shallow sandy bottoms. These fish can actually swim within the sand. *Range*: Indo-Pacific; Cabo San Lucas to Panama and Islas Galapagos.

144. **SOCORRO RAZORFISH**
(cuchillo de Socorro)
***Novaculichthys* sp.**
Identification: This razorfish may be undescribed. The juveniles and females have a grey-white body. with a red-brown stripe along the eye. *Size*: Length to about 3 inches (8 cm). *Habitat*: Coarse sand bottom in depths of 50 to about 80 ft. (15 to 25 m). *Range*: I have observed this razorfish only at Isla Socorro, Mexico.

female above male below

FAMILY SCARIDAE
Parrotfishes

145. **STAREYE PARROTFISH**
(loro de Carolina)
Calotomus carolinus
Identification: The pinkish bands radiating from the eye and unfused teeth are good characters to identify this parrotfish, particularly the males. Females have mottled orange-brown bodies, and the body scales are outlined in white or yellow. *Size*: Length to about 20 inches (50 cm). *Habitat*: Around coral and rock reefs. *Range*: Indo-Pacific; on this coast from Islas Revillagigedo to Islas Galapagos.

146. **LOOSETOOTH PARROTFISH**

Nicholsina denticulata

Identification: This small drab-colored parrotfish also has teeth that are not fused. The color is usually a mottled brown or grey. *Size*: Length to about 8 inches (20 cm). *Habitat*: Algae covered shallow reefs. *Range*: Gulf of California to Ecuador and Islas Galapagos.

147. **AZURE PARROTFISH (perico)**

Scarus compressus

Identification: Males have bright green bodies; each scale is outlined with orange; green streaks radiate from the eye. Juveniles of both sexes are reddish-brown. Adult females are light blue to blue-gray. The posterior edge of the tail fin is straight. *Size*: Length to 2 ft. (0.6 m). *Habitat*: Occur around shallow reefs to depths of about 75 ft. (23 m). *Range*: Bahia Concepcion; central Gulf of California to Islas Galapagos.

148. **BLUECHIN PARROTFISH (loro banba azul)** ***Scarus ghobban***

Identification: Females have light blue bands on a light orange-brown body. Males are bluish-green with a blue chin. The caudal fin is straight with longer upper and lower fin rays. *Size*: Length to 1.5 ft. (0.5 m). *Range*: Indo-Pacific; on this coast from Rocas Alijos; Gulf of California (Isla Carmen) to Islas Galapagos.

female above male below

149. **BUMPHEAD PARROTFISH**

(loro jorabado) ***Scarus perrico***

Identification: The best field characters for adults are the large hump on the head and the rounded posterior edge of the tail fin. *Size*: Length to 2.5 ft. (76.2 cm). *Habitat*: Occurs on rocky reefs to depths of about 100 ft. (30 m). *Range*: Bahia Magdalena, Baja California, and central Gulf (Isla Carmen) to Peru, and most offshore islands. *Natural History*: All parrotfish feed on various species of algae. They "sleep" at night under ledges or in crevices; some species can produce a mucus tent or cocoon. All parrotfish are pelagic spawners.

female above male below

150. **BICOLOR PARROTFISH**

(loro violacio) ***Scarus rubroviolaceus***

Identification: Both adults are bicolored: males are dark green anteriorly and light green posteriorly; females are brownish-red anteriorly and light maroon to tan posteriorly. The posterior profile of the caudal fin is square, but the dorsal and ventral rays of the caudal fin extend beyond the rear edge. *Size*: Length to about 28 inches (70 cm). *Habitat*: This parrotfish prefers reefs with abundant coral beds; found to depths of about 100 ft (30 m). *Range*: Tropical Indo-Pacific; on this coast, central Gulf of California (Isla Monserrate) to Islas Galapagos and most offshore islands.

FAMILY OPISTOGNATHIDAE

Jawfishes

151. **FINESPOTTED JAWFISH**

(boca grande manchada)

Opistognathus punctatus

Identification: This largest of the Gulf of California jawfishes can be identified by the fine dark spots on the head and large dark spots on body. *Size*: Length to 1.3 ft. (0.4 m). *Habitat*: Jawfish construct burrows in shallow, sandy bottoms. *Range*: Bahia Magdalena and the Gulf of California to Panama. *Natural History*: Jawfish incubate their eggs in their mouths. They feed on small bottom dwelling as well as planktonic invertebrates.

152. **BLUESPOTTED JAWFISH (boca manchada azul)**

Opistognathus rosenblatti

Identification: The bright blue spots on the beige to yellow-brown body are very distinctive. *Size*: Length to about 4 inches (102 mm). *Habitat*: In deep burrows in coarse sand, shell fragments, and pebbles at base of reefs. In depths of 15 to 80 ft. (4.5 to 24 m). *Range*: Gulf of California. *Natural History*: These jawfish often form large colonies of up to 100 fish or more. Their burrows are spaced about 3 ft. (1 m) apart.

FAMILY TRIPTERYGIIDAE
Triplefin Blennies

153. **LIZARD TRIPLEFIN (largartija tres aletas)**

Crocodilichthys gracilis

Identification: This most common of the Gulf triplefins can be distinguished by the three dorsal fins, and there is a red and white stripe along the mid-body. The black mark on the caudal peduncle is preceded by a white bar. *Size*: Length to about 2.5 inches (6.4 cm). *Habitat*: On reefs near crevices and under ledges; to depths of about 125 ft. (38 m). *Range*: Gulf of California.

FAMILY LABRISOMIDAE
Clinid Blennies

154. **REDSIDE BENNEY (trambollo rojo)**

Malacoctenus hubbsi

Identification: The pointed snout, the reddish marks on sides of the male, and the brownish broken stripes on the sides of juveniles and females are distinctive. *Size*: Length to 3.5 inches (8.9 cm). *Habitat*: On shallow reefs to depths of 25 ft. (8 m). *Range*: Bahia Sebastian Vizcaino and Gulf of California to Acapulco. *Natural History*: Clinid blennies feed on a variety of small invertebrates such as crabs, snails, chitons, brittle stars, urchins, and worms.

155. **LARGEMOUTH BLENNY (chalapo)**

Labrisomus xanti

Identification: The large head and mouth, and the bright red throat of males during breeding season, are good field characters. *Size*: Length to 7 inches (17.8 cm). *Habitat*: In and around crevices on shallow reefs. *Range*: Bahia Sebastian Vizcaino and Gulf of California to Bahia Tenacatita, Jalisco, Mexico.

FAMILY CHAENOPSIDAE
Tube Blennies

156. **BROWNCHEEK BLENNY (carillo moreno)**

Acanthemblemaria crockeri

Identification: The large dark brown spot that covers most of the gill cover is very distinctive. *Size*: Length to about 2.5 inches (6 cm). *Habitat*: Usually found in empty worm and mollusc tubes. *Range*: Known only from the Gulf of California. *Natural History*: This tube-dweller spends less time in its hole than other tube blennies; often it is seen swimming out in the open, feeding on plankton, small bottom-living invertebrates and even small fishes. The males guard the eggs that the female deposits on the walls of the tube.

157. **ORANGETHROAT PIKEBLENNY (trambollito lucio)**

Chaenopsis alepidota

Identification: The best field characters are the elongate body and the large mouth, i.e., the gape extends behind eye. *Size*: Length to 6 inches (15.2 cm). *Habitat*: Found in parchment worm tubes in sandy, shelly bottoms near reefs; to depths of about 75 ft. (23 m). *Range*: Anacapa Island, California, to Gulf of California.

FAMILY BLENNIDAE
Combtooth Blennies

158. **PANAMIC FANGED BLENNY (trambollito negro)** ***Ophioblennius steindachneri***

Identification: The red ring around the eye, the dark spot behind the eye, and the blunt head are distinctive. *Size*: Length to about 10 inches (25 cm). *Habitat*: Occur on reefs in shallow waters to depths of about 35 ft. (11 m). *Range*: Isla Guadalupe, Baja California, and upper Gulf of California to Peru and offshore Islands. *Natural History*: These very common blennies feed on algae and invertebrates. The females deposit eggs in crevices.

159. **SABERTOOTH BLENNY (diente sable)** ***Plagiotremus azaleus***

Identification: The elongate yellow body, with two dark stripes, and a mouth located below and posterior to the snout are good field characters. *Size*: Length to 4 inches (10.2 cm). *Habitat*: Occur singly and in pairs with aggregations of rainbow wrasse, which their color pattern mimics, around reefs; to depths of about 50 ft. (15 m). *Range*: Gulf of California to Peru, including offshore islands. *Natural History*: These blennies' coloration mimics that of rainbow wrasse (# 141), thus allowing them to approach fish that are expecting to be cleaned of parasites, but instead this blenny takes a bite out of their skin.

FAMILY GOBIIDAE
Gobies

160. **SONORA GOBY (gobio chiquito)** ***Gobiosoma chiquita***

Identification: This drab-colored goby has seven to nine dark bands on the sides. These bands are often in the shape of an hourglass. The center of each band has a dark spot. *Size*: Length to about 2.5 inches (54 mm). *Habitat*: Shallow rock and sandy areas from the intertidal to about 30 ft. (9 m). *Range*: This goby is found only in the Gulf of California. *Natural History*: These daytime foragers feed on small crustaceans and molluscs. The males guard the eggs. Some gobies obtain all or part of their food by picking parasites off other fishes. Other food includes: crabs, shrimps, copepods, molluscs, worms, sponges and eggs of invertebrates and other fishes.

161. **REDLIGHT GOBY (gobio semaforo)**

Coryphopterus urospilus

Identification: The dark spot at the base of the tail and the reddish spots on the whitish body identify this goby. *Size*: Length to 2.5 inches (6.4 cm). *Habitat*: Usually found on sand, around base of reefs to depths of about 125 ft. (38 m). *Range*: Bahia Magdalena and the Gulf of California south to Colombia and Islas Galapagos.

M. Conlin

162. **BANDED CLEANER GOBY (gobio barbero)**

Elacatinus digueti

Identification: This goby lacks scales, the head is bright orange-red, with 3 or 4 bands. There are about 8 or 9 narrow bands on the yellow body. *Size*: Length to about 1.25 inches (32 mm). *Habitat*: In and arond reef crevices. *Range*: Gulf of California to Colombia. *Natural History*: This bright-colored goby is most often seen on or near Panamic green morays, where they forage on or clean parasites from the moray.

A. Kerstitch

163. **WIDEBANDED CLEANER GOBY (gobio)** ***Elacatinus* sp.**

Identification: The orange-red head and dark bands on the body separated by light spaces that are about the same width as the dark bands distinguishes this goby from the banded cleaner goby, *Elacatinus digueti* (# 162), which has wavy dark bands that are narrower than the light interspaces. *Size*: Length to 1.3 inches (3.3 cm). *Habitat*: In rocky crevices, where it cleans parasites from large fish such as moray eels and groupers; to depths of about 50 ft. (15 m). *Range*: Isla Angel de la Guarda in the upper Gulf of California south to Colombia.

164. **REDHEAD GOBY**
(gobio de cabeza rojo)
Elacatinus puncticulatus

Identification: The orange-red head of this goby has two black stripes, with a yellow stripe in between, extending posteriorly from the eye to at least the second dorsal fin. *Size*: Length to 1.8 inches (4.6 cm). *Habitat*: Occurs on reefs, in crevices, and among sea urchin spines; found to depths of about 25 ft. (8 m). *Range*: Upper Gulf of California to Ecuador. *Natural History*: Occasionally these gobies feed on parasites attached to other fish.

165. **BLUEBANDED GOBY**
(gobio bonito) ***Lythrypnus dalli***

Identification: The color pattern of blue bands on the red body is very distinctive. *Size*: Length to 2.2 inches (5.6 cm). *Habitat*: This goby occurs around crevices, under ledges, and among sea urchin spines; found to depths of about 250 ft. (76 m). *Range*: Monterey, California, to Gulf of California and Ecuador.

FAMILY ACANTHURIDAE
Surgeonfishes

166. **WHITECHEEK SURGEONFISH**
(navajon cariblanco)
Acanthurus nigricans

Identification: The best field character is the bright yellow stripes along the base of the dorsal and anal fins. *Size*: Length to about 8.25 inches (21 cm). *Habitat*: Individuals occur around shallow reefs. *Range*: Tropical Pacific; on this coast from Cabo Pulmo in the Gulf of California to Islas Galapagos. *Natural History*: Surgeonfish usually form aggregations, particularly when spawning, which takes place near the surface. The members of this family feed primar-ily on algae or plankton.

167. **CONVICT TANG**
(navajon carcelario)

Acanthurus triostegus

Identification: Six dark bars on a creamy-yellowish or white body are very distinctive. *Size*: Length to about 10.5 inches (26.3 cm). *Habitat*: Around shallow reefs, both alone and in small aggregations. *Range*: Tropical Indo-Pacific; on this coast from Bahia Los Frailes to Panama and offshore islands.

168. **PURPLE SURGEONFISH**
(navajon aleta amarilla)

Acanthurus xanthopterus

Identification: This surgeonfish has a wide yellow stripe that runs through the eye; outer portion of pectoral fins are yellowish. The overall color ranges from dark brown to purplish grey. *Size*: Length to about 24 inches (61 cm). *Habitat*: Around reefs and over sand bottoms, usually up in water column during the day, in depths of 40 to about 300 ft. (12 to 90 m). *Range*: Central Gulf of California to Colombia and Islas Galapagos and other offshore islands. *Natural History*: Purple surgeonfish feed on algae, diatoms and hydroids.

169. **BLUESPOTTED SURGEONFISH**
(navajon estriado)

Ctenochaetus marginatus

Identification: The blue spots on the dark body and whitish stripes on the dorsal and anal fins are very distinctive. *Size*: Length to about 8 inches (20 cm). *Habitat*: Reefs. *Range*: Indo-Pacific; on this coast from Islas Revillagigedo to Panama.

170. **CHANCO SURGEONFISH (navajon barbero)**

Prionurus laticlavius

Identification: This surgeonfish is almost identical to the yellowtail surgeonfish (# 171), but lacks the dense black spots that cover that fish's body. However, the greyish body does usually have a few large black spots near the tail. *Size*: Length to about 24 inches (60 cm). *Habitat*: Around reefs that support various species of algae. Shallow depths to about 50 ft. (15 m). *Range*: Rocas Alijos, off southern Baja California, to Panama and Islas Galapagos and other offshore islands.

171. **YELLOWTAIL SURGEONFISH (cochinito)** ***Prionurus punctatus***

Identification: The yellow tail and gray body, covered with black spots, is easily seen underwater. This surgeonfish can be distinguished from the very similar *P. laticlavius* (# 170), the common yellowtail surgeonfish found around Islas Revillagigedo and Galapagos which lack black spots on most of the body. *Size*: Length to about 2 ft. (0.6 m). *Habitat*: In groups around shallow reefs; to depths of about 100 ft. (30 m). *Range*: Upper Gulf of California to El Salvador.

FAMILY ZANCLIDAE
Moorish Idols

172. **MOORISH IDOL (idolo moro)** ***Zanclus canescens***

Identification: The long snout, long filament on the dorsal fin, black vertical bands on body, and caudal fin are very distinctive. *Size*: Length to about 8 inches (20 cm). *Habitat*: These beautiful fish are usually found in pairs, around shallow reefs. *Range*: Tropical Indo-West Pacific; on this coast from Isla Santa Catalina in Gulf of California to Islas Galapagos.

M. Shargel

M. Shargel

FAMILY SCOMBRIDAE
Mackerels and Tunas

173. WAHOO
(peto) ***Acanthocybium solandri***

Identification: This very popular game fish is readily identified by the long slender body with bluish bands and the long spiny dorsal fin. *Size*: Length to about 65 inches (210 cm). *Habitat*: Surface waters in all tropical oceans, usually more common around sea mounts and offshore islands. *Range*: Southern Baja California to Peru and all oceanic islands. *Natural History*: A single female can spawn as many as 6 million eggs. The eggs are pelagic. These voracious predators feed on a wide variety of fishes.

FAMILY BOTHIDAE
Lefteye Flounders

174. FLOWERY FLOUNDER
(lenguado tropical) ***Bothus mancus***

Identification: This distinctive flounder has a brown body covered with white and pale blue spots, some of which form partial circular patterns. *Size*: Length to about 16.5 inches (42 cm). *Habitat*: Usually observed on reefs or sandy areas adjacent to reefs. *Range*: Indo-Pacific; on this coast from Islas Revillagigedo to Islas Galapagos.

FAMILY PLEURONECTIDAE
Righteye Flounders

175. OCELLATED TURBOT
(platija ocelada)

Pleuronichthys ocellatus

Identification: This flatfish can be identified by one or two ocellated spots on the lateral line at mid-body. *Size*: Length to about 18 inches (45 cm). *Habitat*: Shallow sandy and rocky areas. *Range*: Bahia Magdalena and upper and central Gulf of California.

FAMILY BALISTIDAE
Triggerfishes

176. **FINESCALE TRIGGERFISH**
(peje puerco coche) ***Balistes polylepis***

Identification: The deep body, along with its drab coloration, distinguish this triggerfish. *Size*: Length to 2.5 ft. (0.8 m). *Habitat*: Occurs in aggregations and singly around reefs as well as over sandy bottoms to depths of about 100 ft. (30 m). Occasionally found at night lying on their sides. *Range*: Pt. Saint George, Del Norte County, California, and Gulf of California south to Chile. *Natural History*: These triggerfish feed on sand-dwelling worms and crustaceans as well as attached rocky habitat invertebrates. Triggerfishes use their strong dorsal spines to wedge themselves into crevices at night. They build nests on rubble bottoms; at this time females can be dangerous to divers by giving painful bites in order to drive off the intruder.

T. Uno

177. **BLACK DURGON**
(calafate negro) ***Melichthys niger***

Identification: A blue-black body with white stripes at the base of the dorsal and anal fins are the best field characters. *Size*: Length to about 1 ft. (0.3 m). *Habitat*: This uncommon triggerfish occurs around shallow reefs. *Range*: Indo-Pacific; on this coast Rocas Alijos, Baja California, and Gulf of California to Panama and Islas Revillagigedo, Clipperton, Cocos and Malpelo. *Natural History*: This distinctive triggerfish forages on attached algae as well as drift algae in water column.

178. **BLUNTHEAD TRIGGERFISH**
(pejepuenco de piedra)
Pseudobalistes naufragium

Identification: Most easily identified by the dark bars on a bluish-gray to brownish-gray body and by the prominent forehead. *Size*: Length to at least 3 ft. (0.9 m). *Habitat*: Occur around reefs and over sandy bottoms; to depths of about 100 ft. (30 m). *Range*: Bahia San Quintin, Baja California, and in the Gulf of California from Isla San Francisco to Ecuador.

179. **ORANGESIDE TRIGGERFISH**

(cochino) ***Sufflamen verres***

Identification: The brown body with the orange side patch and the shape of the body are distinctive field characters. *Size*: Length to about 1.3 ft. (0.4 m). *Habitat*: Solitary fish occur around reefs and over sand patches. Found to depths of about 100 ft. (30 m). *Range*: Isla Cedros, Baja California, and in the Gulf of California from Isla Carmen to Ecuador, and most offshore islands.

180. **REDTAIL TRIGGERFISH**

(pez puerco) ***Xanthichthys mento***

Identification: The red tail with a bright blue border and the streaks on the head are easily seen underwater. *Size*: Length to about 11 inches (28 cm). *Habitat*: In midwater and near surface around reefs, anchored boats, and other structures; typically occur in large aggregations. *Range*: Tropical Pacific, including Hawaiian Islands; on this coast from Ventura, California, to Islas Galapagos. Very common around Islas Revillagigedo, and other offshore islands.

FAMILY MONACANTHIDAE

Filefishes

181. **SCRAWLED FILEFISH**

(lijatrompa) ***Aluterus scriptus***

Identification: The bright blue-green scrawl marks and spots on the bluish-gray to olive-brown body and the shape of the body identify this fish. *Size*: Length to about 30 inches (75 cm). *Range*: Worldwide tropical seas; on this coast it has been recorded from Rocas Alijos, and central gulf of California to Islas Galapagos.

182. **VAGABOND FILEFISH**

Cantherhines dumerilii

Identification: This fish is often confused with the surgeonfishes because of the sharp forward curving spines on the caudal peduncle. The two pair of blade-like spines at the base of the tail are the best field character. *Size*: Length to about 15 inches (38 cm). *Habitat*: Frequents shallow reefs. *Range*: Indo-Pacific; on this coast from Los Frailes in the Gulf of California to Islas Revillagigedo, Isla Cocos and Colombia.

FAMILY OSTRACIIDAE

Trunkfishes

183. **SPOTTED BOXFISH (pez caja)**

Ostracion meleagris

Identification: The shape of the body, the placement of the dorsal and anal fins, and the different coloration of the male and female are distinctive. *Size*: Length to 7 inches (17.8 cm). *Habitat*: Around shallow reefs. *Range*: Tropical Indo-Pacific; on this coast from Cabo Pulmo in the Gulf of California to Panama. *Natural History*: Boxfishes spawn near the surface at night. Because of their toxic slime, other fishes usually avoid eating them.

female above male below

FAMILY TETRAODONTIDAE

Puffers

184. **WHITE SPOTTED PUFFER**

(botete pintado) ***Arothron hispidus***

Identification: The greenish-brown body is covered with white spots and the base of the pectoral fins is surrounded by a large black patch. *Size*: Length to about 20 inches (50 cm). *Habitat*: Rocky and coral areas and sand areas at base of reefs. *Range*: Indo-Pacific; on this coast from lower Gulf of California to Panama and Islas Galapagos. *Natural History*: These distinctive puffers feed on algae as well as a variety of small invertebrates. Puffers produce a poison that occurs in their tissues, particularly the liver and ovaries.

185. GUINEAFOWL PUFFER
(botete negro, botete de oro)
Arothron meleagris

Identification: The black body with white spots and the occasional golden color phase are good field characters. *Size*: Length to about 1 ft. (0.3 m). *Habitat*: Occurs around shallow reefs. *Range*: Tropical Pacific; on this coast from Isla San Diego in the Gulf of California to Ecuador, including offshore islands.

Golden color phase

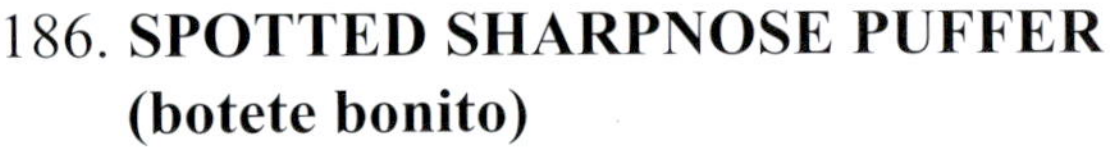

186. SPOTTED SHARPNOSE PUFFER
(botete bonito)
Canthigaster punctatissima

Identification: The large, pointed snout and bluish-white spots on a reddish-brown body are distinguishing field characters. *Size*: Length to 3.5 inches (8.9 cm). *Habitat*: Solitary individuals are found near crevices, caves, and under ledges in shallow waters. *Range*: From Isla Carmen in the Gulf of California south to Panama, including Islas Galapagos and other offshore islands.

187. **BULLSEYE PUFFER (botete diana)**
Sphoeroides annulatus

Identification: The best character is the concentric dark rings on the back. Both the sides and back are also covered with small dark spots. *Size*: Length to about 1.3 ft. (0.4 m). *Habitat*: Shallow, sandy bottoms near reefs and in bays. *Range*: Redondo Beach, California, and the Gulf of California to Peru, including offshore islands.

188. **LOBESKIN PUFFER**
(tamboreta) ***Sphoeroides lobatus***

Identification: The broad head, brown body with white speckles and the skin flaps or lobes along the lower side of the body readily identify this puffer. *Size*: Length to about 12 inches (30 cm). *Habitat*: Usually encountered at base of reefs on sand or rubble bottoms. *Range*: Redondo Beach, California, to Peru and Islas Galapagos.

FAMILY DIODONTIDAE
Porcupinefishes

189. **PACIFIC BURRFISH**
(pejerizo chiquito)
Chilomycterus reticulatus

Identification: This porcupinefish is often confused with the spotted porcupinefish (# 191). The Pacific burrfish has short, fixed spines on the body. *Size*: Length to 22 inches (55 cm). *Habitat*: Around rocky and coral reefs; during the day they are commonly encountered up in the water column. *Range*: Indo-Pacific; on this coast from Long Beach, California, and Gulf of California to Peru, including the offshore islands.

190. **BALLOONFISH**
(pejerizo enmascarado)
Diodon holocanthus

Identification: The balloonfish is most easily recognized by the dark bar over the forehead that extends down past the eye and the long spines that cover the body and head. *Size*: Length to about 1.5 ft. (0.5 m). *Habitat*: Shallow, sandy bottoms and around reefs. *Range*: Tropical seas, worldwide; on this coast from San Mateo County, California, and Gulf of California to Ecuador and offshore islands. *Natural History*: The balloonfish feeds on a variety of sea urchins, molluscs and crabs.

191. **SPOTTED PORCUPINEFISH**
(pez puercoespinas) ***Diodon hystrix***

Identification: This fish differs from the balloonfish by the lack of large blotches and the abundance of black spots on the body, head and tailfin. The spotted porcupine fish differs from the Pacific burrfish (# 189), which has short immovable spines (less than eye diameter) that have three parts at the base. *Size*: Length to 3 ft. (0.9 m). *Habitat*: This nocturnal porcupinefish is more secretive than the balloonfish, staying in caves and crevices during the day and coming out at night to forage on the reef. *Range*: Warm seas of the world; on this coast from San Diego, California, and in the Gulf of California to Chile, including offshore islands.

PICTORIAL KEY TO INVERTEBRATES

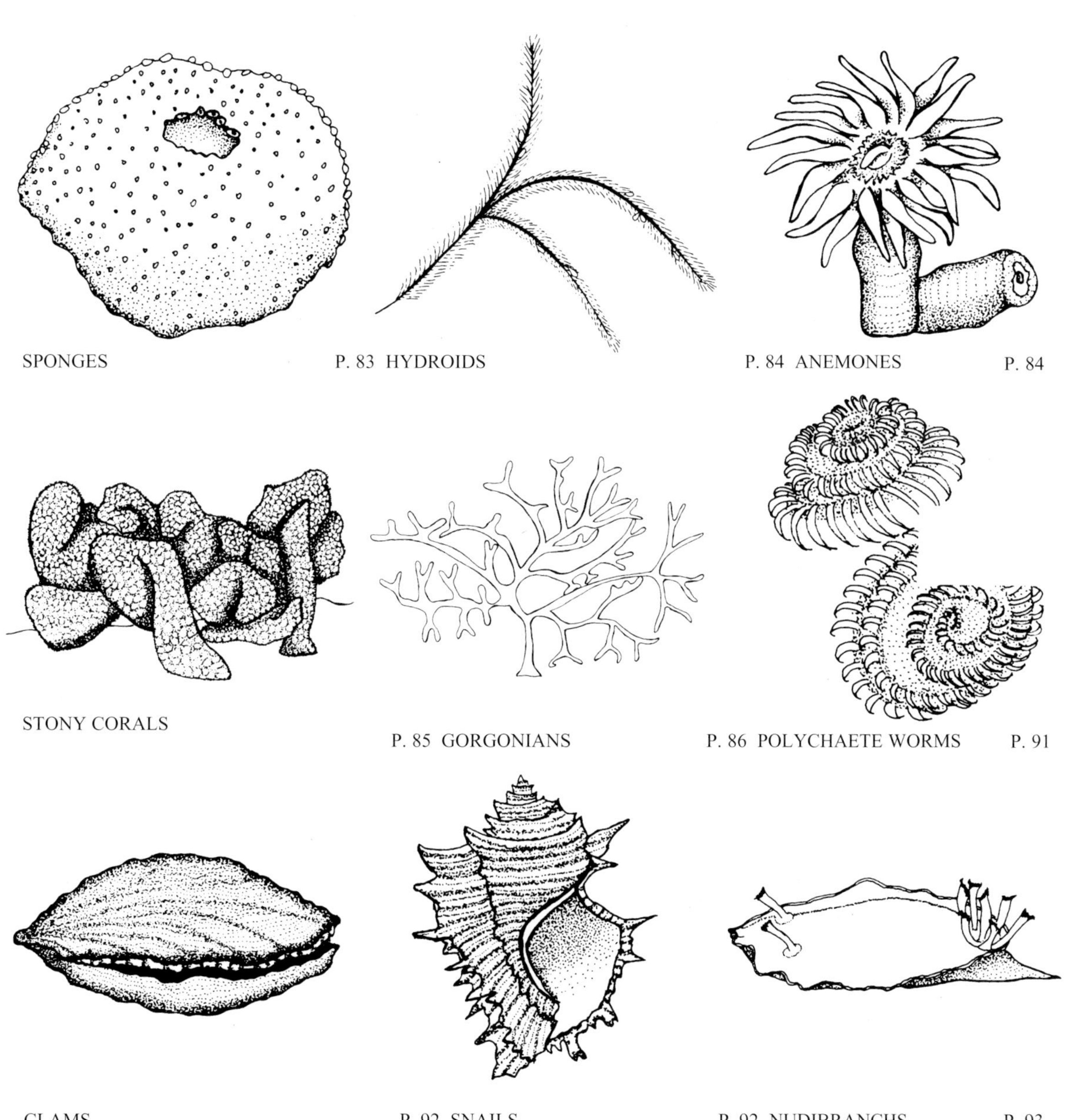

PICTORIAL KEY TO INVERTEBRATES

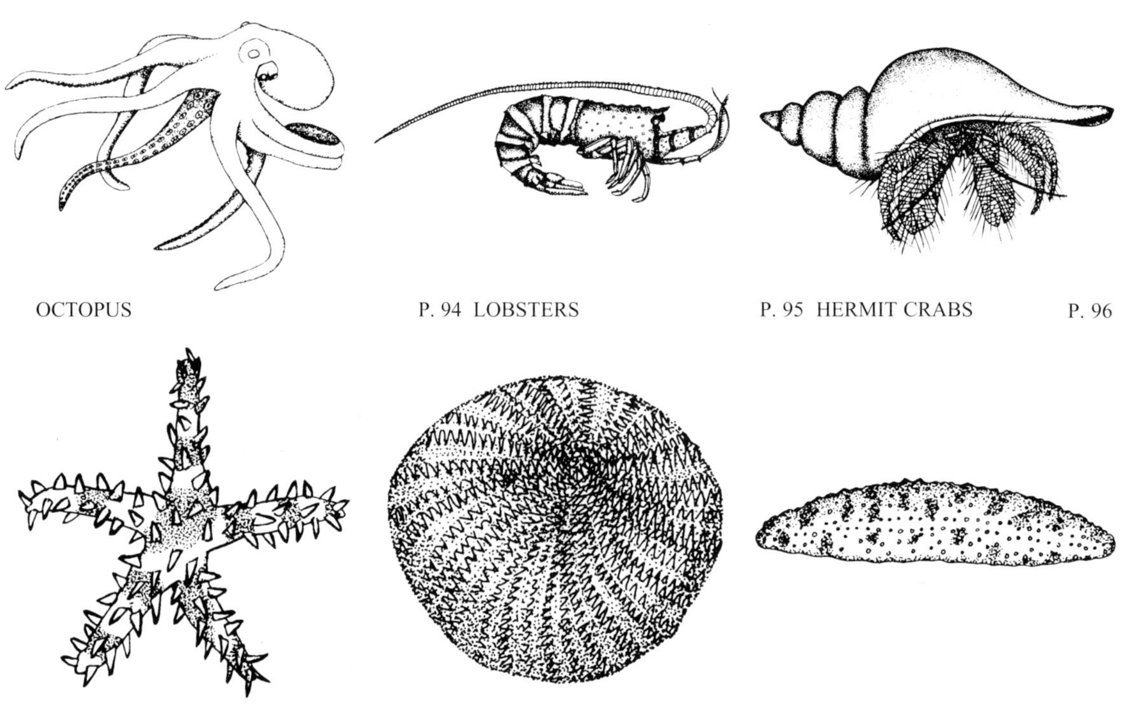

OCTOPUS P. 94 LOBSTERS P. 95 HERMIT CRABS P. 96

97. SEA STARS P. 102 SEA URCHINS P. 103 SEA CUCUMBERS

PHYLUM PORIFERA
Sponges

1. BARREL SPONGE

Pseudosuberites pseudos

Identification: This large sponge is distinguished by an outer surface covered with large tubercles interspersed with craters and pits. There are only a few large excurrent pores (oscula). *Size*: Height to about 1 ft (0.3 m). *Habitat*: These colorful sponges occur on very shallow sand and rocky bottoms. *Range*: Gulf of California; common in Bahia de los Angeles, where they are fed upon by a yellow-gilled porostome nudibranch.

2. SULPHUR SPONGE

Aplysina fistularis

Identification: This common sponge varies greatly in form: two growth forms are illustrated. Color varies from yellow to orange to yellow-orange and brown. *Size*: Height to about 6 inches (15 cm). *Habitat*: Occur on rocks to depths of about 100 ft. (30.5 m). Common in shallow waters of lower Gulf. *Range*: Cosmopolitan; occurs throughout Gulf of California and from Pt. Conception, California, to Cabo San Lucas. *Natural History*: *Tylodina fungina*, an opisthobranch mollusk, feeds on this sponge.

PHYLUM CNIDARIA
Hydroids, Anemones, Corals, Gorgonians

3. STAGHORN HYDROCORAL

Janaria mirabilis

Identification: This hydrocoral usually has three or four symmetrical, upright branches and a horizontal branch. *Size*: Height to about 2 inches (50 mm). *Habitat*: Usually found on soft or rubble bottom in depths of 10 to 500 ft (3-152 m). *Natural History*: The hermit crab, *Manucomplanus varians*, uses this hydrocoral for its home.

4. STINGING HYDROID

Lytocarpus nuttingi

Identification: The best field character is the white, feather-like form of this hydroid, which may be the same species that occurs off Santa Catalina Island, California. This common subtidal hydroid is capable of inflicting painful stings; divers should avoid touching it with their bare skin. *Size*: Colonies may reach 1 ft (0.3 cm) in height. *Habitat*: Shallow reefs to depths of about 60 ft (18 m). *Range*: Bahia Magdalena, Baja California, and central and lower Gulf of California.

5. BURROWING ANEMONE

Pachycerianthus fimbriatus

Identification: The members of this genus are characterized by the parchment-like tube surrounding the column and the long tentacles. *Size*: Height to about 15 inches (37.5 cm). *Habitat*: Sand around reefs to depths of about 150 ft (46 m). *Range*: Alaska to Gulf of California.

6. **SAND ANEMONE**

Alicia beebei

Identification: The long column and tentacles are the best field characters for this species when foraging at night (a). During daylight these anemones contract and withdraw their tentacles (b). *Size*: Height to about 8 inches (20cm). *Habitat*: Shallow sandy bottoms. *Range*: Central Gulf of California to Panama.

7. **RED EPIZOANTHID**

***Epizoanthus* sp.**

Identification: The columns of these colonial anemones are usually covered with sand or mud. The short tentacles are red. *Size*: Height to about 1 inch (25 mm). *Habitat*: Form encrusting growths on alcyonarians or vertical rock surfaces, in depths to about 250 ft (76 m). *Range*: Outer coast of Baja California and upper and central Gulf of California.

8. **ENCRUSTING STONY CORAL**

Porites panamensis

Identification: The hard encrusting form and greenish color of this common stony coral are the best field characters. *Size*: Diameter of colonies about 3ft. (90cm). *Habitat*: Unpolluted, shallow, rocky reefs. *Range*: Gulf of California to Panama. *Natural History*: Crown of thorns seastars feed on these and other hard corals.

9. ELEGANT CORAL

Pocillopora elegans

Identification: This massive coral, consisting of many thick branches with rounded ends and a very rough surface, is very distinctive. The color varies from brown to green. *Size*: Height to about 3.5ft (1m). *Habitat*: Shallow, unpolluted reef in depths of 3 to 50 ft (1 m - 17 m). *Range*: Central Gulf of California to Ecuador, including offshore islands.

10. ORANGE CUP CORAL

Tubastraea coccinea

Identification: The bright yellow to orange color of the tentacles and the deep circular shape of the individual cups (corallites) are good field identification characters. *Size*: Diameter of colonies to about 12 inches (30 cm). *Habitat*: Occur usually on underside of rocks and in crevices and caves; to depths of about 75 ft (23 m). *Range*: Central Gulf of California to Ecuador, including offshore islands.

11. CONSAG CUP CORAL

Bathycyathus consagensis

Identification: These cup corals are distinguished by the brown color and the six to seven large septa, with many smaller septa in between; diameter of the cup exceeds the height. *Size*: Cup diameter to about 37 inches (9 mm). *Habitat*: Shallow reefs to depths of about 325 ft (100 m). *Range*: Rocas Alijos, Baja California, and Gulf of California to Islas Galapagos.

12. **GORGONIAN**

Muricea sp.

Identification: The branches of this gorgonian form almost round clumps rather than branching in a single plane. The polyps are white. *Size*: Height of colony to about 10 inches (25 cm). *Habitat*: On vertical rock faces. *Range*: Bahia Magdelena and Gulf of California.

13. **GORGONIAN**

Muricea appressa

Identification: The yellow polyps of this single plane branching gorgonian are the best characteristic. *Size*: Height to about 30 inches (77 cm). *Habitat*: Shallow reefs to depths of about 100 ft (30 m). *Range*: Gulf of California.

14. **ROBUST GORGONIAN**

Muricea californica

Identification: The thick brown to red branches with white polyps are distinctive. *Size*: Height of colonies to about 32 inches (80 cm). *Habitat*: Low intertidal reefs to about 180 ft (55 m). *Range*: Pt. Conception, California, to Gulf of California and Panama.

15. **BUMPY ORANGE GORGONIAN**

Eugorgia aurantica

Identification: The best field characteristics are the very bumpy orange and yellow branches. The side branches are usually very short. *Size*: Height of colony to about 2.5 ft (0.8 m). *Habitat*: Offshore reefs and islands where these animals can find clean, plankton-rich water; to depths of about 100ft (30 m). *Range*: Isla Cedros on outer coast of Baja California and entire Gulf of California.

16. **YELLOW-POLYP BLACK CORAL**

Antipathes galapagensis

Identification: The bright yellow nonretractable polyps on the slender branching skeleton are distinctive. *Size*: Height to about 6 ft (2 m). *Habitat*: Attached to reefs, wrecks, and other objects in depths to 250 ft (76 m). *Range*: Thetis Bank, Baja California, and central Gulf of California to Ecuador.

17. **RED GORGONIAN**

Eugorgia daniana

Identification: This gorgonian is distinguished by the slender red branches with white polyps. *Size*: Height to about 3 ft (0.9 m). *Habitat*: Offshore reefs and islands to depths of about 100 ft (30 m). *Range*: Central and lower Gulf of California.

18. **YELLOW GORGONIAN**

Eugorgia ampla

Identification: The best field characters are the yellow branches that, in some cases, are attached to their neighbor, forming a partial web. *Size*: Height to about 3 ft (0.9 m). *Habitat*: Offshore reefs and islands; to depths of about 100 ft (30 m). *Range*: Gulf of California.

19. **WHITE GORGONIAN**

Lophogorgia alba

Identification: The long slender white stalks lacking side branches are very distinctive. The polyps are also white. *Size*: Height to 3 ft (0.9 m). *Habitat*: Shallow reefs; to depths of 165 ft. *Range*: Gulf of California to Panama, common in Bahia de los Angeles.

20. **BROWN SEAFAN**

Pacifigorgia sp.

Identification: The red or brown, close-knit connecting branches with white polyps are good field characteristics. *Size*: Height to about 2.5 ft (0.8m). *Habitat*: Offshore rocks and reefs to depths of at least 75 ft (23 m). *Range*: Central and lower Gulf of California.

21. **GORGONIAN**

Adelagorgia phyllosclera

Identification: These sea fans have bright yellow and red branches. The polyps are light yellow. *Size*: Height of colonies to about 3 ft (90 cm). *Habitat*: Shallow, low profile reefs. *Range*: La Jolla, California, to Baja California.

22. **YELLOW SEAFAN**

Pacifigorgia sp.

Identification: The best field characters are the yellow, widely spaced connecting branches with white polyps. *Size*: Height to about 2 ft. (0.6 m). *Habitat*: Deep reefs and submarine canyons; to depths of at least 100 ft. (30 m). *Range*: Lower Gulf of California.

23. **RED SEAFAN**

Gorgonia adamsi

Identification: The red, interconnected, widely spaced branches (spaces are about two branch widths wide) with white polyps are very distinctive. *Size*: Height to about 1 ft. (30 cm). *Habitat*: Offshore reefs and rocks; to depths of about 120 ft. (37 m). *Range*: Gulf of California.

24. **YELLOW GORGONIAN**

Filigella mitsukurii

Identification: The whitish stalks and yellow polyps are good field characters. *Size*: Height to about 1.3 ft. (0.4 m). *Habitat*: Found on sand bottom to depths of at least 100 ft. (30 m). *Range*: Originally described from Japan, on this coast, outer Baja California.

25. **SEA PEN**

Ptilosarcus undulatus

Identification: The best field characters are the large, fleshy lobes on the thick central stalk. The color varies from translucent, light yellow, to red-orange. *Size*: Height to about 1 ft (0.3 m). *Habitat*: Muddy, sandy bottoms to depths of 165 ft. (50 m). *Range*: Throughuot the Gulf of California.

PHYLUM ANNELIDA
Worms

26. **GIANT SPIRALLED POLYCHAETE**

Spirobranchus giganteus

Identification: The two spiralled feeding plumes are very distinctive. The color ranges from white to blue to red-orange. *Size*: Height of plumes to about 1 inch (2.5) cm). *Habitat*: Tubes are attached to coral heads and rocks; found to depths below 200 ft. (61 m). *Range*: Circumtropical; on this coast central and lower Gulf of California to Panama.

27. **PANAMIC FANWORM**

Bispira rugosa monterea

Identification: The white and brown banded feeding plumes are the best field character. *Size*: Height above substrate to about 2 inches (5 cm). *Habitat*: Reefs and sand and gravel bottoms to depths of about 200 ft. (61 m). *Range*: Gulf of California to Panama.

PHYLUM MOLLUSCA
Oysters, Scallops, Clams, Snails Nudibranchs, Octopus, and Squid

28. **PURPLE LIP ROCK OYSTER**

Spondylus calcifer

Identification: The black and white banded fleshy mantle with yellow streaks is the best field character. *Size*: Diameter to about 10 inches (25 cm). *Habitat*: Attached to shallow rocks and reefs to depths of about 75 ft. (23 m). *Range*: Gulf of California to Peru.

29. **MUREX**

Muricanthus princeps

Identification: The spines and the brown-black and white bands are distinctive. *Size*: Length to about 5 inches (13 cm). *Habitat*: Shallow reefs to depths of about 75 ft. (23 m). *Range*: Gulf of California to Peru.

30. **PANAMIC HORSE CONCH**

Pleuroploca princeps

Identification: This snail can be distinguished by the high spire and the orange to brown shell. The foot is orange or red with iridescent blue spots. *Size*: Length to about 18 inches (45 cm). *Habitat*: Around shallow reefs on sand, low intertidal to about 200 ft. (61 m). *Range*: Gulf of California to Peru. *Natural History*: This is the largest snail in the region. They prey on other snails.

31. **MEXICAN DANCER**

Tridachia diomedea

Identification: The best field character for this sea slug is the colorful, fleshy frills on the back and the black-and-yellow lines on the body and rhinophores. *Size*: Length to 2 inches (50 mm). *Habitat*: Mangroves and shallow reefs and rocks; to depths of about 75 ft. (23 m). *Range*: Gulf of California to Panama.

32. **CLOWN NUDIBRANCH**

Chromodoris norrisi

Identification: The purple and yellow dots on the white dorsal surface are distinctive. *Size*: Length to about 2.5 inches (6 cm). *Habitat*: Shallow rocks and algae. *Range*: Isla Cedros, Baja California, and Gulf of California.

33. **SEDNA**

Glossodoris sedna

Identification: The best field characters are the white body with the red tipped gills and rhinophores (the paired fleshy sensory flaps on the head) and the yellow and red margins of the foot. *Size*: Length to 2.5 inches (63 mm). *Habitat*: Tidepools and rocky areas to depths of 65 ft. (19.8 m). *Range*: Gulf of California to Ecuador.

34. **GALACTIC SEA SLUG**

Chromodoris galexorum

Identification: The large red spots with yellow borders are good field characters. Gills and rhinophores are dark red. *Size*: Length to about 1 inch (2.5 cm). *Habitat*: Shallow rocky areas to depths of about 150 ft. (46 m). *Range*: Central and southern Gulf of California.

35. **OCTOPUS**

Octopus sp.

Identification: There are at least nine species of octopus recorded from the Gulf. In most cases, it requires close, in-hand examination to determine the species. The octopus illustrated is the common species I have observed in the waters around Islas Revillagigedo, and best fits the description of *O. hubbsorum*. *Size*: Length to about 2 ft. (61 cm). *Habitat*: Shallow rocky and coral reefs. *Range*: Cabo San Lucas and Islas Revillagigedo.

36. SOCORRO SPINY LOBSTER
Panulirus penicillatus

Identification: The bright blue bases of the antennae are good field characters. *Size*: Weight to about 10 lb. (4.5 kg). *Habitat*: Shallow rock areas; to depths of about 75 ft. (23 m). *Range*: Islas Revillagigedo, Tres Marias, Clipperton, and Galapagos on this coast. Also occurs off South Africa, in the Red Sea, and Indian Ocean.

37. PINTO SPINY LOBSTER
Panulirus inflatus

Identification: The orange spines on the front of the carapace and at the bases of the bluish antennae are distinctive. *Size*: Weight to about 8 lb. (4 kg). *Habitat*: Rocky crevices and caves from shallow depths to 250 ft. (76 m). *Range*: Bahia Magdalena and Gulf of California to the Gulf of Tehuantepec.

38. CALIFORNIA SPINY LOBSTER
Panulirus interruptus

Identification: The general red coloration of the body is the best field character. *Size*: Length to over 2 ft. (60 cm), and weight to 15 lb. (7 kg). *Habitat*: Rocky areas to depths of 250 ft. (76 m). *Range*: Monterey, California, to Bahia Magdalena, and upper Gulf of California.

39. SLIPPER LOBSTER

Scyllarides astori

Identification: The wide, flat body and lack of long antennae are good field characters. *Size*: Length to about 18 inches (45 cm). *Habitat*: Reefs with caves and crevices and sandy bottoms to depths of at least 300 ft. (19 m). *Range*: Upper Gulf of California to Ecuador.

40. HERMIT CRAB

Aniculus elegans

Identification: This large, hairy hermit crab can best be distinguished by the red bands on the legs and chelipeds (claws). *Size*: Length to about 6 inches (15 cm). *Habitat*: Rock, gravel, and coral heads; to depths of 150 ft. (46 m). *Range*: Outer coast of Baja California, Gulf of California to Ecuador.

41. PANAMIC ARROW CRAB

Stenorhynchus debilis

Identification: The reddish-brown triangular-shaped carapace and long legs and chelae are distinctive. *Size*: Carapace length to about 1 inch (25 cm). *Habitat*: In or near reef crevices. *Range*: Santa Catalina Island, California, south to Chile, including the Gulf of California.

PHYLUM ECHINODERMATA
Sea Stars, Brittle Stars, Urchins, Cucumbers

42. SPINY SAND STAR

Astropecten armatus

Identification: There are plates and spines on the edges of each arm of this grey to tan-colored star. The tube feet lack suckers. *Size*: Diameter to about 14 inches (36 cm). *Habitat*: Sand and mud bottoms, from the intertidal to at least 525 ft. (160 m). *Range*: Monterey, California, to Gulf of California and Peru. *Natural History*: These stars feed on molluscs and other soft-bottom dwelling invertebrates. They are also scavengers.

43. CROWN OF THORNS

Acanthaster ellisii

Identification: The 10 to 15 short, spiny arms and the strongly pointed long and short spines on the upper body are very distinctive. *Size*: Diameter to about 1.5 ft. (0.5 m). *Habitat*: Shallow reefs where there is an abundance of hard and soft corals; to depths of 150 ft. (46 m). *Range*: Isla San Diego in Gulf of California, to Peru and offshore islands. *Natural History*: Feed on stony corals.

44. GULF SUN STAR

Heliaster kubiniji

Identification: This many-armed star can be separated from the other Gulf sunstar, *H. microbrachius*, by the presence of 19 to 25 arms (rays) as compared with the 30 to 40 arms of *H. microbrachius*. *Size*: Diameter to about 2 ft. (0.6 m). *Habitat*: Shallow reefs and sand bottoms. *Range*: Bahia Magdalena and Gulf of California to Nicaragua.

45. **FRAGILE STAR**

Linckia columbiae

Identification: The best field characters are the gray and red mottling, arms of unequal length, and the nearly round cross section of the arms. *Size*: Diameter to about 6 inches (15 cm). *Habitat*: Shallow rocky areas; to depths of about 24 ft. (7.3 m). *Range*: San Pedro, California, and the upper Gulf of California to Peru.

46. **TAMARISK SEASTAR**

Tamaria stria

Identification: This dark red star has long cylindrical arms. There are rows of pores on each arm. *Size*: Diameter to about 7 inches (174 mm). *Habitat*: Rocky and subtidal bottoms to depths of 165 ft. (50 m). *Range*: Central Gulf of California to Colombia.

47. **YELLOW SPOTTED STAR**

Pharia pyramidata

Identification: The large yellow to orange spots on the arms are the best field character. *Size*: Diameter to about 1 ft. (0.3 m). *Habitat*: Shallow rocks and reefs to depths of about 456 ft. (139 m). *Range*: Gulf of California to Peru, including the Islas Galapagos.

48. **TAN STAR**

Phataria unifascialis

Identification: This star resembles the yellow spotted star, except that the color pattern is tan with two orange stripes on each arm. *Size*: Diameter to about 1 ft. (0.3 m). *Habitat*: Shallow rocks and reefs to depths of about 450 ft (139 m). *Range*: Bahia Magdalena; Bahia de los Angeles in upper Gulf of California to Peru, including the Islas Galapagos.

49. **ORANGE STAR**

Echinaster tenuispina

Identification: The color patterns of yellow-brown to orange or red and the dark red to black spot on the tip of each arm is distinctive. *Size*: Diameter to about 4 inches (10 cm). *Habitat*: Shallow rocks and reefs to depths of 240 ft. (73 m). *Range*: Scammon's Lagoon, Baja California, to Gulf of California. Common only in upper Gulf.

50. **CHOCOLATE CHIP STAR**

Nidorellia armata

Identification: The short, thick arms and color patterns of black spots on the cream colored upper body are very distinctive. *Size*: Diameter to about 5 inches (13 cm). *Habitat*: Shallow reefs, to depths of 240 ft. (73 m). *Range*: Bahia Magdalena, Baja California, and Gulf of California to Peru.

51. **PANAMIC CUSHION STAR**

Pentaceraster cumingi

Identification: The robust arms covered with bright red spines are very distinctive. *Size*: Diameter to about 6 inches (15 cm). *Habitat*: Shallow reefs and sand bottoms to depths of 600 ft. (183 m). *Range*: Gulf of California to Peru, including Islas Galapagos. This is the most abundant star at scuba depths in the central and lower Gulf.

52. **SPINY STAR**

Amphiaster insignis

Identification: The best field characters are the long, robust spines on the red arms and body. *Size*: Diameter to about 7 inches (174 m). *Habitat*: Shallow reefs, sand and mud bottoms to depths of 420 ft. (128 m). *Range*: Central Gulf of California to Panama; uncommon.

53. **BRADLEY'S SEA STAR**

Mithrodia bradleyi

Identification: The long slender arms covered with short spines and the red and brown color are good field characters. *Size*: Diameter to about 14 inches (35 cm). *Habitat*: Shallow reefs to depths of about 165 ft. (50 m). *Range*: Central Gulf of Panama, including Islas Revillagigedos and Galapagos.

54. **KEELED STAR**

Asteropsis carinifera

Identification: The ridge of spines in the center of each arm, the flattened body, and triangular arms are distinctive. *Size*: Diameter to 10 inches (25 cm). *Habitat*: Reefs from intertidal to 120 ft. (37 m). *Range*: Central Gulf of California to Islas Galapagos and Indo-Pacific.

55. **BASKET STAR**

Astrodictyum panamense

Identification: The best field characters are the many branched, brown banded arms. *Size*: Length of extended arms to about 1 ft. (0.3 m). *Habitat*: Found attached to gorgonians; shallow water to depths of 600 ft. (183 m). *Range*: Punta Eugenia, Baja California, and Gulf of California to Panama.

56. **BASKET STAR**

Astrocaneum spinosum

Identification: The single row of white spots, circled by black rings, on the glossy-tan arms is very distinctive. *Size*: Diameter to about 12 inches (0.3 m). *Habitat*: On gorgonians. *Range*: Isla Cedros on outer coast of Baja California and Gulf of California to Panama.

57. SEA URCHIN

Arbacia incisa

Identification: The dark color and lateral spines that are slightly longer than the diameter of the test are good field characters. *Size*: Diameter to about 3 inches (8 cm). *Habitat*: Shallow reefs and rocks. *Range*: Gulf of California to Peru, including the Islas Galapagos.

58. SLATE PENCIL URCHIN

Eucidaris thouarsii

Identification: The thick, blunt-tipped, pencil-like spines are very distinctive. *Size*: Diameter to about 7 inches (18 cm). *Habitat*: Found in crevices on shallow reefs; to depths of about 500 ft. (152 m). *Range*: Santa Catalina Island, California, and the Gulf of California to Ecuador, including the offshore islands.

59. FLOWER URCHIN

Toxopneustes roseus

Identification: The very short spines and long, petal shaped pedicellariae give this urchin a distinctive flower appearance. *Size*: Diameter to about 5 inches (13 cm). *Habitat*: Shallow reefs, sand and mud bottoms to depths of 80 ft. (24 m). *Range*: Bahia Tortugas, Baja California, and central Gulf of California to Ecuador, including offshore islands.

60. **BROWN URCHIN**

Tripneustes depressus

Identification: The abundant, short spines and large test are very distinctive. *Size*: Diameter to about 7 inches (18 cm). *Habitat*: Shallow reefs. *Range*: La Paz to Islas Galapagos, including Islas Revillagigedos.

61. **CROWNED URCHIN**

Centrostephanus coronatus

Identification: The very long, slender, serrated spines and dark purple color are good field characters. The crowned sea urchin resembles the toxic urchin, *Diadema mexicanum* (NI), but the latter species has long, smooth, white and purple banded spines. *Size*: Diameter to about 6 inches (15 cm), spines about twice as long as diameter of test. *Habitat*: Shallow reefs, in crevices; to depths of about 350 ft. (107 m). *Range*: Channel Islands, California, and Gulf of California to Ecuador and the offshore islands.

62. **CUCUMBER**

Holothuria zacae

Identification: The best field characters are the many short papillae on the dark gray body with only a few large papillae. *Size*: Length to about 1 ft. (30 cm). *Habitat*: Shallow reefs and sand and mud bottoms. *Range*: Santa Catalina Island, California, and the Gulf of California to Islas Galapagos.

63. BROWN CUCUMBER

Isostichopus fuscus

Identification: This cucumber has a chocolate-brown to orange-brown, heavy, rigid body with large, orange papillae on the back. Skin smooth to the touch. Underside of body is flat, with three rows of feet. *Size*: Length to 12 inches (30 cm). *Habitat*: Reefs and sand bottoms in shallow water to depths of 200 ft. (61 m). *Range*: Bahia Vizcaino, Baja California, and upper Gulf of California to Ecuador and the Islas Galapagos.

64. BROWN SPOTTED CUCUMBER

Holothuria impatiens

Identification: The brown spotted body covered with many black-tipped papillae is distinctive. *Size*: Length to about 6 inches (15 cm). *Habitat*: Shallow sandy and mud bottoms and rocky areas to depths of 150 ft (46 m). *Range*: Bahia Rosario, Baja California, and Gulf of California to Ecuador and Islas Galapagos.

H. Bertsch

65. SYNAPTED CUCUMBER

Euapta godeffroyi

Identification: The best field characters are the long, soft, snake-like body that lacks tube feet and the long feather-like feeding tentacles. *Size*: Length when fully extended to at least 3 ft (0.9 m). *Habitat*: These snake-like cucumbers come out to feed at night on and around shallow rocks and reefs to depths of 150 ft (46 m). *Range*: Tropical Indo-Pacific; on this coast from the central gulf to Panama.

BIBLIOGRAPHY

Allen, G. R. and D. R. Robertson. 1994. *Fishes of the Tropical Eastern Pacific.* University of Hawaii Press, Honolulu, Hawaii. 332 p.

Behrens, D. W. 1991. *Pacific Coast Nudibranchs, A Guide to the Opisthobranchs of the Northeastern Pacific.* 2nd Edition. Sea Challengers, Monterey, California. vi + 107 p.

Brusca, R. C. 1980. *Common Intertidal Invertebrates of the Gulf of California.* University of Arizona Press, Tucson, Arizona. 513 p.

Gotshall, D. W. 1989. *Pacific Coast Inshore Fishes.* 3rd Edition. Sea Challengers, Monterey, California. 96 p.

Gotshall, D. W. 1994. *Guide to Marine Invertebrates - Alaska to Baja California.* Sea Challengers, Monterey, California. vi + 105 p.

Hobson, E. S. 1968. *Predatory behavior of some shore fishes in the Gulf of California.* U.S. Fish and Wildlife Service Research Department. 73: 92 p.

Miller, D. W., and R. N. Lea. 1976. *Guide to the Coastal Marine Fishes of California.* Department of Fish and Game, Fish Bulletin (157): 1-249 p.

Grove, J. S. and R. J. Lavenberg. 1997. *The Fishes of the Galapagos Islands.* Stanford University Press, Stanford, California. 861 p.

Kerstitch, A. N. 1989. *Sea of Cortez Marine Invertebrates - A Guide for the Pacific Coast, Mexico to Ecuador.* Sea Challengers, Monterey, California. v + 114 p.

Morris, R. H., D. P. Abbott, and E. C. Haderlie. 1980. *Intertidal Invertebrates of California.* Stanford University Press, Stanford, California. 690 p.

Randall, J. E. 1996. *Shore Fishes of Hawaii.* Natural World Press, Vida, Oregon. 216 p.

Robins, C. R., R. M. Bailey, C. E. Bond, J. R. Brooker, E. A. Lachner, R. N. Lea and W. B. Scott. 1991. *Common and Scientific Names of Fishes from the United States and Canada.* 5th Edition. American Fisheries Society, Bethesda, Maryland. 183 p.

Thomson, D. A., L. T. Findley, and A. N. Kerstitch. 1979. *Reef Fishes of the Sea of Cortez.* University of Arizona Press, Tucson, Arizona. 302 p.

Thomson, D. A., and N. McKibbin. 1976. *Gulf of California Fishwatchers Guide.* Golden Puffer Press, Tucson, Arizona. 75 p.

Walford, L. A. 1974. *Marine Game Fishes of the Pacific Coast from Alaska to the Equator.* Smithsonian Institution Reprint by T.F.H. Publications, Inc., Neptune, New Jerssey. 205 p. 70 plates.

INDEX

A

Abudefduf troschelii 52
Acanthaster ellisii 97
Acanthemblemario crockeri 68
Acantocybium solandri 78
Acanthuridae 71
Acanthurus nigricans 71
Acanthurus triostegus 72
Acanthurus xanthopterus 72
Acapulco damselfish 54
Adelagorgia phyllosclera 90
Alicia beebei 85
Almaco jack 38
Alphestes immaculatus 26
Alphestes multiguttatus 26
Aluterus scriptus 76
Amphiaster insignis 100
Anemones 84
Angelfishes 50
Aniculus elegans 96
Anisotremus davidsonii 42
Anisotremus interruptus 42
Anisotremus taeniatus 43
Annelida 91
Antennariidae 22
Antennarious sanguineus 22
Antipathes galapapensis 88
Aplysina fistularis 83
Apogon atricaudus 33
Apogon guadalupensis 33
Apogon pacifici 34
Apogon retrosella 34
Apogonidae 33
Arbacia incisa 102
Argus moray 19
Arothron hispidus 77
Arothron meleagris 78
Arthropoda 95
Asteropsis carinifera 101
Astrocaneum spinosum 101
Astrodictyum panamense 101
Astropecten armatus 97
Aulostomidae 24
Aulostomus chinensis 24
Azure parrotfish 65
Azurina hirundo 52

B

Balistes polylepis 75
Balistidae 75
Balloonfish 80
Banded cleaner goby 70
Banded guitarfish 15
Banded lizardfish 22
Barberfish 50
Barracuda 58
Barred pargo 39
Barred serrano 31
Barrel sponge 83
Barspot cardinalfish 34
Basket star 101
Bathycyathus consagensis 86
Beaubrummel 55
Belonidae 23
Bicolor parrotfish 66
Bigeye jack 36
Bigeyes 32
Bigscale soldierfish 23
Bispira rugosa monterea 92
Black durgon 75
Blackfin soapfish 32
Blennidae 69
Bloody frogfish 22
Blue-and-gold snapper 41
Blue-and-yellow chromis 53
Bluebanded goby 71
Blue-bronze chub 47
Bluechin parrotfish 65
Blue-spotted jack 36
Bluespooted jawfish 67
Blunthead triggerfish 75
Bodianus diplotaenia 58,59
Bradley's seastar 100
Bothidae 74
Bothus mancus 74
Broomtail grouper 29
Browncheek blenny 68
Brown cucumber 104
Brown sea fan 89
Brown spotted cucumber 104
Brown urchin 103
Bullseye electric ray 15
Bullseye puffer 79
Bullseye stingray 16
Bumphead damselfish 53
Bumphead parrotfish 66
Bumpy orange gorgonian 88
Butterflyfishes 49
Burrito grunt 42
Burrowing anemone 84

C

Calamus brachysomus 45
California needlefish 22
California sheephead 62
California spiny lobster 95
Calotomus carolinus 64
Cantherhines dummerilii 77
Canthigaster punctatissima 78
Carangidae 35
Caranx orthogrammus 35
Caranx caballus 35
Caranx caninus 36
Caranx lugubris 36
Caranx melamphygus 36
Caranx sexfasciatus 36
Carcharhinidae 13
Carcharhinus falciformis 13
Carcharhinus galapagensis 13
Cardinalfishes 33
Cephalopholis panamensis 28
Centrostephanus coronatus 103
Chaenopsidae 68
Chaenopsis alepidota 68
Chaetodipterus zonatus 48
Chaetodontidae 49
Chaetodon humeralis 49
Chamleon wrasse 60
Chanco surgeonfish 73
Chilomycterus reticulatus 79
Chocolate chip star 99
Chromis alta 52
Chromis atrilobata 53
Chromis limbaughi 53
Chromodoris galexorum 94
Chromodoris norrisi 93
Cirrhitichthys oxycephalus 57
Cirrhitidae 57
Cirrhitus rivulatus 57
Clarion angelfish 50
Clarion damselfish 56
Clarion soldierfish 23
Clinid blennies 67
Clown nudibranch 93
Clupeidae 21
Combtooth blennies 69
Cnidaria 84
Conger eels 21
Congridae 21
Consag cup coral 86
Convict tang 72
Coral hawkfish 57
Corals 84

INDEX

Cornetfishes ... 24
Cortez angelfish ... 51
Cortez chub ... 47
Cortez damselfish ... 54,55,56
Cortez garden eel ... 21
Cortez grunt ... 43
Cortez rainbow wrasse ... 63
Cortez round stingray ... 17
Cortez soapfish ... 32
Coryphopterus urospilus ... 70
Crabs ... 96
Croakers ... 45
Crocodilichthys gracilis ... 67
Crowned urchin ... 103
Crown of thorns ... 97
Ctenochaetus marginatus ... 72
Cucumbers ... 103
Cup corals ... 86

D

Damselfishes ... 52
Darkblotch scorpionfish ... 25
Dasyatidae ... 16
Dasyatis brevis ... 16
Dermatolepsis dermatolepis ... 27
Diadema mexicanum ... 103
Diamond stingray ... 16
Diodon holocanthus ... 80
Diodon hystrix ... 80
Diodontidae ... 79
Diplectrum pacificum ... 26
Diplobatis ommata ... 15

E

Eagle rays ... 17
Echeneidae ... 35
Echinaster tenuispina ... 99
Echinodermata ... 97
Elacatinus digueti ... 70
Elacatinus puncticulatus ... 71
Elacatinus sp. ... 70
Elagatis bipinnulata ... 37
Electric rays ... 15
Elegant coral ... 86
Elopidae ... 21
Elops affinis ... 21
Encrusting stony coral ... 85
Ephippidae ... 48
Epinephelus analogus ... 27
Epinephelus itajara ... 27
Epinephelus labriformis ... 28
Epizoanthus sp. ... 85
Euapta godeffroyi ... 104
Eucidaris thouarsii ... 102
Eugorgia ampla ... 89
Eugorgia aurantica ... 88
Eugorgia daniana ... 88

F

Filefishes ... 76
Filigella mitsukurii ... 91
Finespotted jawfish ... 66
Fine-spotted moray ... 18
Finescale triggerfish ... 75
Fistularia commersonii ... 24
Fistulariidae ... 24
Flag cabrilla ... 28
Flagtail blanquillo ... 34
Flower urchin ... 102
Flowery flounder ... 74
Forcipiger flavissimus ... 49
Fragile star ... 98
Froghishes ... 22

G

Gafftopsail pompano ... 38
Galactic Sea slug ... 94
Galapagos shark ... 13
Gerreidae ... 42
Gerres cinereus ... 42
Giant damselfish ... 54
Giant electric ray ... 15
Giant hawkfish ... 57
Giant spiralled polychaete ... 91
Girella simplicidens ... 46
Glasseye ... 32
Glossodoris sedna ... 94
Gnathanodon speciosus ... 37
Goatfish ... 46
Gobies ... 69
Gobiidae ... 69
Gobiosoma chiquita ... 69
Golden grouper ... 30
Golden jack ... 37
Golden snapper ... 41
Gold-spotted sandbass ... 31
Goliath grouper ... 27
Gorgonia adamsi ... 90
Gorgonian ... 87
Grammistidae ... 32
Graybar grunt ... 44
Green jack ... 35
Green wrasse ... 63
Grunts ... 42
Guadalupe cardinalfish ... 33
Guineafowl puffer ... 78
Guitarfishes ... 14
Gulf grouper ... 29
Gulf opaleye ... 46
Gulf sun star ... 97
Gymmnomuraena zebra ... 18
Gymnothorax castaneus ... 18
Gymnothorax dovi ... 18

H

Haemulidae ... 42
Haemulon flaviguttatum ... 43
Haemulon maculicauda ... 43
Haemulon sexfasciatum ... 44
Haemulon steindachneri ... 44
Halichoeres chierchiae ... 59
Halichoeres dispilus ... 60
Halichoeres insularis ... 60
Halichoeres nicholsi ... 61
Halichoeres semicinctus ... 61
Hammerhead sharks ... 14
Harengula thrissna ... 21
Hawkfishes ... 57
Heliaster microbrachius ... 97
Heliaster kubiniji ... 97
Hermit crab ... 96
Hermosilla azurea ... 47
Herrings ... 21
Heteroconger diqueti ... 21
Heteropriacanthus cruentatus ... 32
Hippocampus ingens ... 25
Holacanthus clarionensis ... 50
Holacanthus passer ... 51
Holocentridae ... 23
Holothuria impatiens ... 104
Holothuria zacae ... 103
Hoplopagrus guntheri ... 39
Hourglass moray ... 19
Hydroids ... 84

I

Island jack ... 35
Island wrasse ... 63
Indigo wrasse ... 60
Isostichopus fuscus ... 104

J

Jacks ... 35
Jawfishes ... 66
Jewel moray ... 19

INDEX

Johnrandallia nigrirostris 50

K
Keeled star 101
King angelfish 51
Kyphosidae 46
Kyphosus analogus 47
Kyphosus elegans 47
Kyphosus lutescens 48

L
Labridae 58
Labrisomidae 67
Labrisomus xanti 68
Largemouth blenny 68
Latin grunt 44
Leather bass 27
Lefteye flounders 74
Leopard grouper 30
Linckia columbiae 98
Liopropoma fasciatum 28
Lizardfishes 22
Lizard triplefin 67
Lobeskin puffer 79
Lobsters 95
Longnore butterflyfish 49
Longnose hawkfish 57
Loosetooth parrotfish 65
Lophogorgia alba 89
Lutjanidae 39
Lutjanus aratus 40
Lutganus argentiventris 40
Lutjanus guttatus 40
Lutjanus inermis 41
Lutjanus novemfasciatus 41
Lutjanus viridis 41
Lythrypnus dalli 71
Lytocarpus nuttingi 84

M
Machete 21
Mackerels 74
Malacanthidae 34
Malacanthus brevirostris 34
Malacoctenus hubbsi 67
Manta birostris 17
Manta rays 17
Melichthys niger 75
Mexican barracuda 58
Mexican dancer 93
Mexican goatfish 46
Mexican hogfish 58,59
Mexican lookdown 37
Microlepidotus inornatus 44
Microspathodon bairdi 53
Microspathodon dorsalis 54
Mithrodia bradleyi 100
Mobulidae 17
Mojarras 42
Mollusca 92
Monacanthidae 76
Moorish idol 73
Morays 18
Mugil sp. 58
Mugilidae 58
Mullet 58
Mullet snapper 40
Mullidae 46
Mulloidichthys dentatus 46
Muraena argus 19
Muraena clepsydra 19
Muraena lentiginosa 19
Muraenidae 18
Murex 92
Muricanthus princeps 92
Muricea appressa 87
Muricea californica 87
Muricea sp. 87
Mycteropterca jordani 29
Mycteroperca prionura 29
Mycteroperca rosacea 30
Mycteroperca xenarcha 29
Myliobatidae 17
Myrichthys tigrinus 20
Myripristis clarionensis 23
Myripristis berndti 23
Myripristis leiognathos 23

N
Narcine entemedor 15
Needlefishes 22
Nematistiidae 39
Nematistius pectoralis 39
Nicholsina denticulata 65
Nidorellia armata 99
Novaculicthys sp. 64
Novaculichtys taeniorus 62
Nudibranchs 93

O
Ocellated turbot 74
Octopus 94
Octopus hubbsorum 94
Octopus sp. 94
Odontoscion xanthops 45
Ophichthidae 20
Ophichthus triserialis 20
Ophioblennius steindachneri 69
Opistognathidae 66
Opistognathus punctatus 66
Opistognathus rosenblatti 67
Orange cup coral 86
Orangeside triggerfish 76
Orange star 99
Orangethroat pikeblenny 68
Ostraciidae 77
Ostracion meleagris 77
Oxycirrhites typus 57
Oysters 92

P
Pachycerianthus fimbriatus 84
Pacific burrfish 79
Pacific creolefish 30
Pacific crevalle jack 36
Pacific dog snapper 41
Pacific flatiron herring 21
Pacific porgy 45
Pacific sandperch 26
Pacific seahorse 25
Pacific snake eel 20
Pacific spadefish 48
Pacifigorgia sp. 89,90
Panama graysby 28
Panamic arrow crab 96
Panamic cushion star 100
Panamic fanged blenny 69
Panamic fanworm 92
Panamic green moray 18
Panamic horse conch 93
Panamic porkfish 43
Panamic sergeant major 52
Panamic soldierfish 23
Panulirus inflatus 95
Panulirus interruptus 95
Panulirus penicillatus 95
Paralabrax auroguttatus 31
Paralabrax maculatofasciatus 31
Paranthias colonus 30
Pareques viola 46
Parrotfishes 64

INDEX

Peacock wrasse 64
Pentaceraster cumingi 100
Pharia pyramidata 98
Phataria unifascialis 99
Pink cardinalfish 34
Pinto spiny lobster 95
Plagiotremus azaleus 69
Plain cardinalfish 33
Pleuronectidae 74
Pleuronichthys ocellatus 74
Pleuroploca princeps 93
Pocillopora elegans 86
Pomacanthidae 50
Pomacanthus zonipectus 51
Pomacentridae 52
Popeye catalufa 33
Porcupinefishes 79
Porgies 45
Porifera 83
Porites panamensis 85
Priacanthidae 32
Prionurus laticlavius 73
Prionurus punctatus 73
Pristigenys serrula 33
Pseudobalistes nauforgium 75
Pseudojulis sp. 60
Pseudosuberites pseudos 83
Ptilosarcus undulatus 91
Puffers 77
Purple lip rock oyster 92
Purple surgeonfish 72

R

Rainbow basslet 28
Rainbow chub 48
Rainbow runner 37
Rainbow scorpionfish 26
Red epizoanthid 85
Red gorgonian 88
Redhead goby 71
Redlight goby 70
Red sea fan 90
Redside blenny 67
Redtail triggerfish 76
Reef cornetfish 24
Remora 35
Remora remora 35
Requiem sharks 13
Rhincodontidae 13
Rhinobatidae 14
Rhincodon typus 13
Rhinobatus productus 14
Righteye flounders 74
Robust gorponian 87
Rock croaker 46
Rockmover wrasse 62
Rock wrasse 61
Roosterfish 39
Round stingray 16
Rypticus bicolor 32
Rypticus nigripinnis 32

S

Sabertooth blenny 69
Salema 45
Sand anemone 85
Sargo 42
Sargocentron suborbitalis 24
Sawtail grouper 29
Scalloped hammerhead 14
Scaridae 64
Scarus compressus 65
Scarus ghobban 65
Scarus perrico 66
Scarus rubroviolaceus 66
Scianidae 45
Scissortail damselfish 53
Scombridae 74
Scorpaena guttata 25
Scorpaena histrio 25
Scorpaena mystes 25
Scorpaenidae 25
Scorpaenodes xyris 26
Scorpionfishes 25
Scrawled filefish 76
Scuticaria tigrina 20
Scyllarides astori 96
Scythe butterflyfish 49
Sea basses 26
Sea chubs 46
Sea fans 90
Seahorse 25
Sea pen 91
Sea stars 97
Sea urchin 102
Sectator ocyurus 48
Sedna 94
Selene brevoorti 37
Seriola lalandi 38
Seriola rivoliana 38
Serranidae 26
Serranus psittacinus 31
Serranus socorroensis 32
Shovelnose guitarfish 14
Silky shark 13
Silverstripe chromis 52
Slate pencil urchin 102
Slipper lobster 96
Snails 92
Snake eels 20
Snappers 39
Soapfish 32
Socorro chub 48
Socorro razorfish 64
Socorro serrano 32
Socorro wrasse 60
Sonora goby 69
Spadefishes 48
Sparidae 45
Sphoeroides annulatus 79
Sphoeroides lobatus 79
Sphyraena lucasana 58
Sphyraenidae 58
Sphyrna lewini 14
Sphyrnidae 14
Spinster wrasse 61
Spiny sand star 97
Spiny star 100
Spirobranchus giganteus 91
Spondylus calcifer 92
Spongers 83
Spotted eagle ray 17
Spottail grunt 43
Spotted boxfish 77
Spotted cabrilla 27
Spotted porcupinefish 80
Spotted rose snapper 40
Spotted sand bass 31
Spotted sharpnose puffer 78
Squirrelfishes 23
Stareye parrotfish 64
Steel pompano 39
Stegastes acapulcoensis 54
Stegastes flavilatus 55
Stegastes leucorus 55
Stegastes rectifraenum 56
Stegastes redemptus 56
Stenorhynchus debilis 96
Stereolepis gigas 27
Stinging hydroid 84
Stingrays 16
Stone scorpionfish 25
Stony corals 85
Strongylura exilis 22

INDEX

Stufflamen verres 76
Sulphur sponge 83
Surgeonfishes 71
Swallowtail damselfish 52
Synapted cucumber 104
Syngnathidae 25
Synodontidae 22
Synodus lacertinus 22

T
Tamarisk sea star 98
Tamaria stria 98
Tan star 99
Tetraodontidae 77
Thalassoma grammaticum 63
Thalassoma lucasanum 63
Thalassoma virens 63
Threebanded butterflyfish 49
Tiger reef eel 20
Tiger snake eel 20
Tilefishes 34
Tinssel squirrelfish 24
Torpedinidae 15
Toxopneustes roseus 102
Trachinotus rhodopus 38
Trachinotus stilbe 39
Triaenodon obesus 14
Tridachia diomedea 93
Triggerfishes 75
Triplefin blennies 67
Tripneustes depressus 103
Tripterygiidae 67
Trumpetfishes 24
Trunkfishes 77
Tubastraea coccinea 86
Tube blennies 68

U
Urchins 102
Urolophidae 16
Urolophus concentricus 16
Urolophus halleri 16
Urolophus maculatus 17

V
Vagabond filefish 77

W
Wahoo 74
Wavyline grunt 44
Whale shark 13
Widebanded cleaner goby 70
Whitecheek surgeonfish 71
White gorgonian 89
White spotted puffer 77
Whitetail damselfish 55
Whitetip reef shark 14
Worms 91
Wounded wrasse 59
Wrasses 58

X
Xenistius californiensis 45
Xyrichthys pavo 64

Y
Yelloweye croaker 45
Yellowfin mojarra 42
Yellow gorgonian 89,91
Yellow-polyp black coral 88
Yellow sea fan 90
Yellow snapper 40
Yellow spotted star 98
Yellowtail 38
Yellowtail surgeonfish 73

Z
Zanclidae 73
Zanclus canescens 73
Zapteryx exasperata 15
Zebra moray 18
Zebra perch 47